VGM Opportunities Series

# OPPORTUNITIES IN
# ELECTRICAL TRADES

# Robert Wood

Revised by
**Kenneth R. Edwards**
Director of Research and Technical Services
International Brotherhood of Electrical Workers

Foreword by
**J. J. Barry**
International President
International Brotherhood of Electrical Workers

**VGM Career Horizons**
*NTC/Contemporary Publishing Group*

**Library of Congress Cataloging-in-Publication Data**

Wood, Robert B., 1934–
    Opportunities in electrical trades / Robert Wood ; rev. by Kenneth
R. Edwards ; foreword by J. J. Barry.
        p.    cm. — (VGM opportunities series)
    Includes bibliographical references.
    ISBN 0-8442-4666-2 (hardcover : alk. paper)
    ISBN 0-8442-4667-0 (pbk. : alk. paper)
    1. Electric engineering—Vocational guidance.    I. Edwards,
Kenneth R.    II. Title.    III. Series.
TK159.W66    1996
621.3'023—dc20                                                    96-26665
                                                                          CIP

HF
5381
.0676
n.5

**Cover photo credits:**

Upper left courtesy of Centel; upper right courtesy of Manpower;
lower left and lower right courtesy of the International Brotherhood of
Electrical Workers.

Published by VGM Career Horizons
A division of NTC/Contemporary Publishing Group, Inc.
4255 West Touhy Avenue, Lincolnwood (Chicago), Illinois 60646-1975 U.S.A.
Copyright © 1997 by NTC/Contemporary Publishing Group, Inc.
Printed in the United States of America
International Standard Book Number: 0-8442-4666-2 (cloth)
                                                  0-8442-4667-0 (paper)
18    17    16    15    14    13    12    11    10    9    8    7    6    5    4    3    2

# CONTENTS

An overview of the electrical industry. A brief history
of the industry. Current trends and future projections.
Earnings and general working conditions. Desirable
attributes and assets.

Education and training. Employment services. Assessing
your own abilities. Seeking the services of a guidance
counselor.

Prerequisite courses. Industrial arts. Vocational and
industrial education. Sample apprenticeship schedules.
Apprenticeship training.

Construction electrician. Maintenance electrician. The
electric motor shop journeyman. Marine electrician. Utility
and power plant occupations. Electric sign serviceperson.

# ABOUT THE AUTHOR

Robert Wood has been a member of the International Brotherhood of Electrical Workers for over 35 years. His electrical background encompasses experience in the following areas: electrical construction, maintenance, electric motor repair and armature rewinding, marine and ship repair, and industrial electronics.

Mr. Wood completed his apprenticeship training by working on the job during the day and attending related classes at night. Later, he returned to night school and earned certification in the field of electronics.

Upon completion of his apprenticeship, Mr. Wood worked for a number of years as a journeyman electrician, and on several occasions was employed in supervisory positions. After acquiring several years of experience at the journeyman level (and passing the required examination), he received a Class A Electrician's License for the city of New Orleans.

Mr. Wood's background also includes several years of teaching experience. He was a member of the faculty of Delgado College of New Orleans, where he taught at both the trade school and junior college levels. During his teaching tenure, he taught skill-improvement courses at night to journeymen electricians and developed a laboratory manual for use in an electrical machinery course.

A member of the United States Coast Guard Reserve, Mr. Wood has often served as instructor in marine and shipboard electrical systems. He holds a commission in the Coast Guard Reserve, where has has served as educational and training officer.

In addition to the diplomas he received for his studies of electricity and electronics, Mr. Wood was awarded an associate degree in electrical engineering from Delgado College and a bachelor of science degree in industrial education from Northwestern State College of Louisiana. He has also studied at the University of Virginia.

Mr. Wood now serves as a member of the Labor Research Advisory Council of the United States Department of Labor, Bureau of Labor Statistics, and the Labor Sector Advisory Committee for Trade Negotiations, as well as the United Way's Environmental Scan Committee.

This edition of *Opportunities in Electrical Trades Careers* was revised by Kenneth R. Edwards, the Director of Research and Technical Services at the International Brotherhood of Electrical Workers. Mr. Edwards has over 40 years of experience in the electrical and electronic industries. He has received numerous awards for his work in education.

# ACKNOWLEDGMENTS

I wish to thank my wife, Joyce, for her perseverance in typing the manuscript of this book, and all my former students who afforded me the opportunity of sharing in a mutual learning process—education. Especially, I would like to tender my appreciation to Kenneth Edwards, Director of Research and Technical Services for the International Brotherhood of Electrical Workers, for his valuable assistance.

I also want to express my gratitude to the following organizations that so willing cooperated in supplying information that added to the value of the publication: the International Brotherhood of Electrical Workers, the National Electrical Contractors Association, the National Electric Sign Association, the Electric Apparatus Service Association, the Edison Electric Institute, the International Association of Electrical Inspectors, the National Fire Protection Association, and the National Electrical Manufacturers Association.

# PREFACE

Thousands of young people will be seeking employment in the electrical industry as we head into the next century. For most of them, the decision to do so will mark a pivotal point in their lives. Consequently, current information is critical in assisting them in making the proper choice.

Occupational selection should be based on many factors. The individual's ability and interests should be weighed heavily. Job potential, the nature of the work, opportunities for advancement, employment security, earnings, and general working conditions are other important elements to consider. Most importantly, rewarding and productive careers depend on being given opportunities and challenges that are in line with an individual's talents.

Prudent career planning depends not only on an individual's assessment of his or her abilities and desires, but also on the advice of professionals and other experienced people. The purpose of this book is twofold—to serve future workers in their decision-making process and to serve as a source of information for those who have a role in career development—counselors, teachers, parents, and counselor-educators.

The book includes topics of interest to individuals contemplating careers in the electrical trades. Beginning with an overview of the electrical industry, the book then covers guidelines for occupational planning and selection, apprising the student of techniques utilized in determining occupational compatibility. Chapter 3 discusses education and training. Selected occupations in the field, opportunities for advancement, and specialization are treated in Chapters 4 through 6. Chapter 7 is statistical in nature, covering such areas of interest as wages and fringe benefits, hours of work, and future work force requirements. The final chapter serves to introduce readers to various organizations in the field.

# FOREWORD

The freedom of individuals to choose their occupations and the type of work that they perform is an article of American faith. In a complex, highly specialized industry, such as the electrical industry, individuals need help to make effective use of this freedom. Particularly during their high school years, young people need assistance to clarify their understanding of their aptitudes, interests, and vocational goals; to learn about the range of occupations, not only in their own but in other communities; to relate their present schooling to their future work and career; and to become informed about the opportunities for further education and training. When a person makes an educational or occupational decision without due consideration of their strengths and opportunities, they waste their potential abilities, and both the community's and nation's human resources are correspondingly weakened. It is with this mind that this book has been written.

The electrical industry is a diverse collection of many branches that includes almost every sector of American society. These sectors include the production and distribution of electrical power and its conversion into other energy forms; the manufacturing of equipment and devices that use this energy; the wiring of homes, offices, and industrial plants; and the use of this energy to supply our comforts and entertainment needs. This diverse industry not only enhances this nation's productivity, but also provides the basis for individual growth.

Reports focusing on America's competitive position suggest that most jobs within the next few years will require more knowledge, training, and higher levels of skills than they do today. The best jobs will call for combining a high level of technical knowledge and high levels of analytical and communication skills. They will need these skills not only because the jobs of the future will require them, but also to prepare the workforce of tomorrow for lifetime education. As technology advances and work forces are downsized, workers will have to return to educational and training systems more often.

In the electrical industry, relatively rapid technological change has always been a fact of life. Research into more efficient ways of generating and applying electricity has always been an important component of our industry. Today we can also add to that two more ingredients: conservation and the quality of electric power.

In looking to the future of electrical construction, we see an era of modernization, maintenance, and retrofitting. Specialty work is becoming more important. This includes advanced computerized control systems; electronic lighting control; factory automation to the point that an auto dealer needing a part can call the factory and have the part made and shipped the same day—with a minimum of worker involvement; photonic systems; advanced telecommunications; and fire and security systems available for every home at reasonable cost. Computer power conditioning and complete energy management will be available for home use. Our entertainment systems will feature the latest movies and concerts brought to us by high resolution digital video and sound transmission and will be available for viewing and listening on wall-size screen panels with surround sound.

Not only will those entering the workforce need new skills but these skills will have to be certifiable. The idea of 12 to 14 years of education will be come a thing of the past for those that seek long-term employment.

The International Brotherhood of Electrical Workers is made up of 750,000 members. These members are employed in more than 770 different occupations in the electrical industry. These members serve your needs in a wide variety of ways. A book describing each and every one of these occupations would be highly impractical and only tend to confuse the reader. Therefore, the author has picked key classifications in several branches of the electrical industry, described them, and forecasted their growth patterns for the future.

J. J. Barry
International President
International Brotherhood
of Electrical Workers

# CHAPTER 1

# A VIEW OF THE FIELD

One advantage that an individual in this country has over individuals in many other countries is the freedom of choice. He or she not only has the right to think and speak freely, but also the freedom to choose an occupation. However, this freedom of occupational choice is sometimes restricted by conditions over which the individual has no immediate control. These conditions are sometimes related to the current economic climate or a given industry or due to the lack of proper preparation. This is especially true of the electrical industry.

With such an enormous diversity of job opportunities in the electrical industry, which range from broadcasting to telecommunications, it is essential for someone desiring to enter the industry to start preparing in high school. No longer can an individual simply accept an assigned curriculum in high school and expect to get a job right after graduation—or after leaving high school without a diploma.

While in high school the individual should make a tentative occupational choice. Occupational choice is a gradual process in which a compromise is reached between a person's abilities, interests, and values and the realistic necessities presented by his or her actual opportunities for preparation and employment. Although knowledge of the world of work is clearly not the only element needed to make sound decisions about one's education, it is highly valuable. Unless you deliberately take the steps needed to supply yourself with such knowledge, you will be at a disadvantage in the job market.

Learning about the world of work and what is required will help motivate you to take full advantage of your educational opportunities. Employers are not only interested in a person's education. They are also interested in any previous work experience, activities in school and in the community, attitudes, attendance in school, driving records, and whether or not a person uses drugs.

You will want to select courses that will open up doors to future educational and occupational choices. All too often, students forget that education can prepare them for a challenging career. In forgetting the importance of education, they often lose out on employment opportunities.

Employment in the electrical industry can be challenging, demanding, and rewarding, because of the range of skills and abilities that are required by the industry. Employment in the electrical industry is not static, in that it is always changing. The knowledge "half life" of someone entering the industry today is five years. Those entering the industry—if they want to be successful and stay employed—will have to make lifelong learning an integral part of their career. For example, in the past 10 years, linemen working for certain electric utilities have had to learn how to use robotic arms and computers, and to work from running boards of helicopters while hovering over energized power lines.

Our nation's labor force is made up of over 125 million workers engaged in a multitude of occupations. In addition to these millions of employed, there are an additional 4.4 million working part time.

The vast and complex economy of the United States offers numerous career choices. Thousands of different types of jobs are available, with a wide range of employers. Moreover, the United States Department of Labor, Bureau of Labor Statistics projects that the total labor force will reach about 147 million by the year 2005. This means that the present labor force will be expanded by about 12.2 percent. This is a much slower rate than the labor force has expanded during the past 10 years.

This book provides the information that will assist those making career choices gain an understanding of the opportunities that the electrical industry has to offer. It also details the educational or skill requirements that the industry demands of those who wish to enter this exciting industry.

## AN OVERVIEW OF THE ELECTRICAL INDUSTRY

The electrical industry is a unique industry, in that it cuts across every segment of this nation's economy. There are over 1000 different types of jobs in the electrical industry. Over 800 of these jobs have some sort of upward mobility. Out of these 800 occupations, over 200 have been recognized by the U.S. Department of Labor's Bureau of Apprenticeship and Training as being apprenticeable.

One segment of this industry is the contract construction industry, which includes establishments engaged primarily in the contract construction of new work, additions, alterations, and repairs. In its Standard Industrial Classification (SIC) Manual, the Office of Management and Budget lists three broad types of contract construction establishments: general building contractors engaged primarily in the construction of residential, farm, industrial, commercial, and public or other buildings (SIC 15); general contractors engaged in such heavy construction as highways and streets, bridges, sewers, railroads, irrigation projects, flood control projects, and marine construction (SIC 16); and special trade contractors who undertake such specialized activities as plumbing, painting, plastering, carpentering, and electrical work (SIC 17).

These specialized activities may be broken down farther into subspecialties for special trade electrical contractors:

1731.01  Electrical contractors
1731.02  Electrical contractors (commercial and industrial work only)
1731.03  Factory maintenance (electrical contractors)
1731.04  Lightning protection equipment contractors
1731.05  Electrical contractors—marine
1731.06  Electronic control systems contractors
1731.07  Telecommunications contractors
1731.08  Telephone installation contractors
1731.09  Landscape lighting contractors
1731.10  Cable wire installation voice and data
1731.11  Electric cable fault locating contractors
1731.12  Pay telephone and booth equipment contractors
1731.13  Telecommunications equipment contractors
1731.14  Time lock contractors
1731.15  Data communications systems contractors
1799.32  Cable splicing contractors
1799.35  Computer room and equipment installation contractors

The construction industry is projected to gain 490,000 jobs over the next 10 years, with an average growth rate of 0.9 percent. The electrical portion of the construction industry is expected to grow to a little over 600,000 by the year 2005. This is an annual growth rate of approximately 1 percent.

## Construction

Electrical construction consists of the installation of electrical wiring, devices, appliances, and equipment. The electrical construction industry is usually separated into two distinct categories: inside construction and outside construction. The inside construction branch can be further subdivided into three segments: residential, commercial, and industrial.

*Residential jobs* consists of electrical and communications wiring, and electrical and electronic devices installed in homes and related types of buildings. *Commercial construction* involves electrical, communications and electronic work installed in commercial establishments, such as supermarkets, shopping centers, drugstores, service stations, and garages. *Industrial work* involves the electrical installations in the construction of petroleum, chemical, textile, rubber, auto, paper, utility and metallurgical plants.

According to Census Bureau figures, there were 63,334 inside electrical contractors in 1995, who employed approximately 560,000 persons. Thirty-four thousand of these contractors employ less than five persons. Labor requirements are proportionate to the size and type of job; whereas residential installations usually do not require more than two workers, the average commercial job generally requires five to eight workers, and large industrial jobs frequently utilize more than fifty journeyman electricians. The annual dollar volume of electrical construction currently exceeds $46 billion, which, with the exception of housing, is mainly for remodeling and repair work.

The construction industry is projected to gain 490,000 jobs over the next 10 years, with an average growth rate of 0.9 percent. The electrical portion of the construction industry is expected to grow to a little over 600,000 by the year 2005. This is an annual growth rate of approximately 1 percent. This slow but steady growth is a result of an increase in the number of households and industrial plants, as well as the desire to upgrade existing buildings. The construction industry as a whole is sensitive to economic fluctuations, so employment will vary from year to year.

Outside construction pertains to the installation of electrical overhead and underground lines and those lines used for communications and cable television. In 1995, there were 1565 contractors doing this type of work. These contractors employed a bit over 20,000 persons. However, over 3,000 persons capable of doing this work were unemployed in 1995. With the downsizing of both electric utility and telephone companies, there will be little growth other than replacement jobs in outside construction for the next few years.

## Electric Utilities

From a functional or operational point of view, one might say that the electrical industry begins with electric utilities. The initial generation of electricity takes place in electric utility plants, providing us with the energy that heats and lights our homes, powers our appliances, factories and industrial plants, and allows us to perform numerous tasks with very little physical effort.

The generation and distribution of electrical power is the primary function of this branch of the electric industry. As our population grows, the demand for housing increases, thereby creating a need for more electricity. Since the latter part of the 1970s, the incremental use of electricity has slowed from its previous growth rate of doubling every eight years. Nonetheless, as our nation grows and our economy expands, electric utilities will obviously need to grow in a commensurate fashion to meet society's needs. Moreover, the share of electricity in the nation's energy profile will continue to expand, rising 9 percent over the next 15 years. The industry currently is experiencing a great deal of downsizing, which has been brought on by deregulation and the growth of independent power producers.

## Maintenance

Electrical maintenance workers service and repair the electrical machines, equipment, and circuits that keep our nation's power plants, factories, hospitals, schools, apartments, and office build-ings operating effectively. This sector of the industry is charged with the responsibility of ensuring safe, steady, and reliable elec-trical service. To accomplish this task, they routinely perform pre-ventative maintenance on electrical equipment and systems. In 1992 maintenance and repair expenditures were $42 billion. This is expected to grow to $45.3 billion by 1998, an increase of roughly 2 percent.

## Specialties

The electrical industry also encompasses a number of sectors that are primarily engaged in performing specialized functions. The electric motor shop industry, marine and shipboard electrical work, and the installation, repair, and service of electrical signs are typical examples of such specialties. The specialized electrical industry has been slowly vanishing. This is largely due to the fact that demand for some of these items has dropped over the years. In some cases, such as small electric motors, it sometimes costs more to repair or rewind a small motor than to buy a new one.

## Telecommunications

Telecommunications are an integral part of the electrical indus-try. It is essentially concerned with the installation, operation, repair, and service of electronic transmitting, and receiving equip-ment. Workers in this field install and maintain communication systems. They install various types of cables capable of carrying hundreds of telephone messages simultaneously—thus providing our country with one of the finest communication networks in the world.

Automation and modern technology have brought a number of innovations to this field. Much of the manual equipment and many of the physical operations once required to perform a specific function have been completely replaced by automated equipment capable of functioning instantaneously and with a high degree of precision.

Fiber-optic and twisted pair technology are now at the forefront in this industry. Fiber-optic systems involve the transmission of light signals over glass fibers. Just as electrical signals are transmitted over copper wire, now light signals can be transmitted over glass filaments. Fiber-optic cable is flexible and lightweight. Cables with 144 optical fibers and the capability of carrying more than 50,000 telephone conversations simultaneously are no bigger around than your finger.

The 1990s have been characterized by record-breaking growth in wireless telecommunication systems. These wireless systems include cellular, paging, satellite, and specialized mobile radio. This thriving market will have enormous growth potential through the year 2000. In the past three years each of these services have had an annual growth rate of 23 to 39 percent. One of the newest fields is direct-to-home satellite broadcasting services. While at present there are only two major carriers, the Digital Satellite System and the PrimeStar system, several other companies plan to be in the market within the next two years.

Cable television (CATV) is another example of a field that is experiencing rapid growth. Cable television makes it possible to bring to the home or office a wider selection of media material than is available by standard broadcast. Cable in the future could replace telephone circuits and offer those subscribing a wide range of interactivity, such as shopping and banking from home.

## Manufacturing

Electrical manufacturing is the key to the electrical industry. Without the manufacturing of electrical products, there would be nothing to use electricity on. Workers employed in manufacturing plants are responsible for the production of an infinite variety of electrical and electronic items that range from simple household fuses to control systems for exploring outer space.

While the segments of the electrical manufacturing industry are still downsizing and will continue to do so well into the next century, certain other segments are expected to grow. Growth is expected to increase 15 to 20 percent in the manufacture wind-energy turbines and other renewable energy equipment.

The combination of these branches and others, such as broadcast, transportation, recording, and entertainment, makes an intricate, dynamic industry that encompasses a multitude of occupations.

## A BRIEF HISTORY OF THE INDUSTRY

During the latter part of the 19th century, many scientists and inventors attempted to produce electricity for practical purposes. Records indicate the occasional use of electric lamps and other devices, such as the telegraph during that period. However, the success of these early products, except for the telegraph, was usually short-lived. It was not until 1879, when Thomas Edison developed his incandescent lamp with a carbonized filament enclosed in an evacuated glass envelope, that the electric lamp began to produce fruitful results. Then, with the advent of the dynamo (now known as the electric generator) and the alternating current system made possible by transformers a system capable of delivering electrical power to densely populated communities emerged.

In the 1880s, commercial uses of electricity began to appear. Although the first generating stations were rather small in capacity, they helped set the stage for future electrification of our cities.

In fact, on September 4, 1882, Thomas Edison announced that he was ready to test a new lighting system that occupied one square mile of New York City. Edison's venture was a success, and thus created the need for a new type of worker.

The duties of early electrical workers pertained primarily to the erection of distribution systems. These systems included the conductors, commonly called "lines," and the poles used to support them. It was from this type of work that the "line person" job classification originated. The line person's work was exciting and interesting, but it was also very dangerous. The fatality rate was high, and insurance was virtually impossible to obtain. The work was hard, the hours were long, and the pay was low. Fortunately, there were enough dedicated, courageous people to lay the foundation upon which today's electrical industry is built.

As electrical power became available to more communities, the need for a larger, more versatile work force grew. While the line-person continued to install and maintain the transmission and distribution systems, a demand developed for electrical workers skilled in wiring residences, buildings, and other commercial establishments. These workers had to be knowledgeable in the art of installing electrical circuits and equipment in a safe, professional manner. Although the tools of the trade were rather simple in those days, workers still had to be trained to use them skillfully. These workers were essentially concerned with wiring buildings, and since a large part of their work was performed inside these buildings, they became known as *inside journeymen*. Today, these highly skilled workers comprise one of the leading groups in the building trades division of the construction industry.

As technology advanced, industry began to appreciate the advantages of electricity for power. Soon, more and more manufacturing firms began to electrify their operations, reinforcing the growth of this branch of the industry.

Electrical manufacturing has contributed a great deal to the industrialization of our nation, and has also made it possible for us to enjoy the benefits of the numerous electrical and electronic devices on the market today. The mass production of appliances and machines has played an important role in providing us with a higher standard of living and more leisure time.

With the country's industrialization came a need for workers to fill the many newly established jobs within the electrical industry. One of the areas that developed rapidly was the service, repair, and maintenance of electrical systems and equipment. The workers in this area soon became known as *maintenance electricians.* The nature of their duties has not changed significantly until recently, when they have been confronted with a much greater degree of sophistication, such as computer networks, advanced communications systems, fire alarm, control schemes, and energy management systems.

Within a short time, electric motor repair and electric sign service began to emerge as distinct specialties, and these two branches have become vital segments of the electrical industry today.

## CURRENT TRENDS AND FUTURE PROJECTIONS

With increased use of electricity in the United States, it is evident that there will be a continual need for skilled workers in the electrical industry. The construction of homes, shopping centers, schools, and other public facilities will remain high to meet the needs of our growing and shifting population. Business expansion and renovation of existing buildings also will require more building and construction work. The development of renewable energy resources and transportation systems will provide for additional jobs in the construction industry. According to the United States

Department of Labor, Bureau of Labor Statistics, employment in the construction sector is expected to increase from about 3.6 million to 4.2 million by the year 2005.

Data released by the Bureau of Labor Statistics indicate that in 1994 there were more than 437,000 construction electricians. This is estimated to increase by 5 percent by the year 2005. In addition to this anticipated growth, thousands of job vacancies will arise from the need to replace journeymen who change their employment to another sector within the field, seek employment in other industries, or retire.

## EARNINGS AND GENERAL WORKING CONDITIONS

Building trades workers receive the highest hourly wages paid to skilled workers, and, within this group, the electrician is quite often the pacemaker. The average weekly wage for a construction electrician in August, 1995 was $677. Table 1 is a comparison between the wages of a construction electrician and some of the other occupations in the construction industry.

**Table 1—Average Weekly Earnings of Selected Construction Trades**
August 1995

| | |
|---|---|
| Plumbing, heating and air conditioning | $632.02 |
| Painting and paper hanging | $529.03 |
| Masonry, stonework, and plastering | $563.84 |
| Carpentry and floor work | $563.60 |
| Roofing, siding, and sheet metal work | $486.72 |

Although a construction electrician may receive a higher hourly wage than other trades, the work in the construction industry is seasonal. Further, because of the nature of construction work, the journeyman usually has limited opportunity to work overtime. The combination of these two factors has had a marked influence on the annual earnings of these workers. Consequently, when one compares the annual earnings of the construction electrician to those of the maintenance electrician, journeyman lineperson, shop journeyman, and electric sign service person, one will note no substantial difference in certain geographical areas.

*Apprentices.* Most apprentices are employed on a specified hourly wage basis, as provided in the collective bargaining agreement. An apprentice's wage is usually expressed as a percentage of the journeyman rate. It ranges from about 35 percent during the first six months of employment to approximately 90 percent as the apprentice enters the last six months of the fifth year in training.

*Workweek.* Generally, the basic workweek consists of 40 hours, the aggregate of a five-day work week of eight hours per day. However, there are a number of workers in the industry who enjoy a standard workweek of 35 hours or less. There is also the tendency for some to work four 10-hour days instead the traditional five day week.

*Premium pay.* On most new construction jobs, electricians who are required to work over 40 hours average per week receive overtime. Some union collective bargaining agreements covering utility, maintenance, and electric motor shop employees specify 1½ times the regular rate of pay for overtime work. Other collective bargaining agreements set the premium rate of pay for overtime at two times the regular hourly rate. Thus, overtime pay rates vary according to the particular branch of the industry, with a range of 1½ to two times the straight time hourly rate.

*Vacations.* Paid leisure is an accepted part of the American way of life, with longer vacations and more paid holidays being enjoyed by a larger segment of the country's work force. Most collective bargaining agreements covering electrical workers provide for paid vacations. The length of the vacation is usually related to the term of employment, especially in the non-construction sectors of the industry. Vacations range from five days for the first year of employment to 30 days for 25 to 30 years of employment.

*Other Fringe Benefits.* Today many workers are covered by health and welfare insurance programs. They also enjoy such benefits as dental and eye care insurance, prescription drug programs, educational assistance, legal assistance, and in many cases some sort of pension plan.

Vacation plans in the construction industry are usually formulated on a cents-per-hour worked basis. For example, assume that an electrician receives 75 cents per hour vacation pay and is employed for 2,000 hours during the course of the year. The amount of vacation pay the worker would be entitled to is the product of $.75 \times 2,000 = \$1,500$, which in terms of time off would be approximately two weeks of paid vacation.

Compensation for holidays not worked is one of the most common fringe benefits provided for organized workers, according to data reported by the Bureau of Labor Statistics. BLS reports that 62 percent of the workers employed by medium and large firms receive ten or more paid holidays. This benefit is not prevalent in the construction industry, but, as the desire for more leisure time gains impetus, it should begin to receive greater acceptance. In short, the electrical industry provides many lucrative opportunities for those seeking careers in this progressive field.

## DESIRABLE ATTRIBUTES AND ASSETS

The electrical trades require a unique combination of talents. Those who enter the field must possess a natural aptitude for using tools and must be able to master the required theoretical discipline. Of even greater importance, the trainee should possess a keen interest and desire for achievement in the field.

## Physical Requirements

Apprenticeship applicants should be physically qualified. That is, they should be able to perform the required manual functions in a safe, professional manner. Although much of their work is done indoors, they should be alert, agile, and physically capable of working outdoors under adverse weather conditions. Applicants for apprenticeship in the electrical construction industry may be required to furnish evidence that they can meet the minimum physical requirements of the job and they are not using any abusive substance.

Being able to drive a lightweight truck or van can be an asset to workers pursuing a career in the electrical field. If the contractor works in more than one state and transports materials or personnel from one state to the other, the individual may be called upon to meet the requirements established by the U.S. Department of Transportation. Knowledge of the area's geography in which you will be working is quite beneficial, especially in terms of understanding street maps and determining effectively how to get from one place to another.

## Age and Education Requirements

Beginning apprentices in the electrical industry cannot be under 16. In the electrical construction industry, the minimum age is 18. A high school education is essential. Applicants will be required to furnish a copy of their high school transcript. Applicants are required to have as a minimum of one year of Algebra with a passing grade. Additional math and science will be regarded as a bonus to the applicant.

Becoming an electrician can be an interesting and challenging assignment, especially for those who enjoy physical as well as mental achievement. In short, a successful career in the electrical field depends largely on the attitude and enthusiasm displayed by the apprentice.

## CHAPTER 2

# PLANNING AND SELECTING
# AN OCCUPATION

Planning a career in the skilled trades involves careful preparation. This preparation must start in early childhood by reading or listening to stories about vocations. When a person reaches the middle school or junior high school years, they should start developing a career plan. This plan does not have to be cast in stone, but should give the individual some idea of those occupations that are growing and the prerequisites that are needed to enter the occupation. By the time one is a junior or senior in high school, it is often too late to get the required subjects that an employer looks for in a job applicant. The successes and failures experienced by many young people are closely correlated to the diligence with which they plan for their vocation. Assuming that you are interested in the electrical field and that you have the potential for meeting the criteria given in Chapter 1, you should become acquainted with the various ways in which you can improve your chances of embarking on a successful career in the electrical trades.

## EDUCATION AND TRAINING

In order to make this country more competitive with world markets, the educational level of the country's work force will be forced to rise during the next century and before. If a person wants to stay employed, education and training will become a lifetime pursuit. Presently, almost every skilled occupation requires its workers to have a high school education. At the turn of the century, many skilled occupations will require addition education and training. The half life of skills in some occupations will be reduces to 18 months from the current five years. While some industries hire trained high school dropouts, the skill standards set now being established by a large number of industries will prevent this in the future.

Education is a highly marketable commodity. It provides the individual with the tools necessary for upward mobility in the workplace. Traditionally, high school graduates have enjoyed substantially greater earnings over their working careers than those who did not complete high school.

In the near future, a certificate of initial mastery will be required to entry the workforce or to continue one's education. While many employers continue to accept a GED as the evidence of a high school education, the requirements for receiving such a certificate are changing, and an increasing number of employers have set higher education requirements in order to maintain a high performance work place.

### GED Tests

In 1945, a program known as General Education Development (GED) was set up to accommodate returning veterans who had not completed their high school training. The American Council on Education developed a GED test battery that covers the following five disciplines:

1. Correctness and effectiveness of expression
2. Interpretation of literary materials
3. General mathematical ability

4.  Interpretation of reading materials in social studies
5.  Interpretation of reading materials in natural sciences

Since its inception, the scope of the program has broadened to the extent that many states now allow anyone 17 years or older to participate. Each state department of education establishes the policies and procedures for an adult resident to earn a high school equivalency diploma based on the results of the GED test. The state department of education determines the minimum test scores required to attain a diploma. It also establishes such criteria as age, residency, and any previous high school enrollment requirements for admission to the program. These regulations vary from state to state. If you are interested in this program, contact your state's department of education for specific details about its administration or write to the American Council on Education, One Dupont Circle, Washington, DC 20036.

One alternative to the GED programs is distant learning, which involves being accepted by a community college and completing 6 to 12 credit hours of instruction in courses available by electronic-based classes, such as the electronic university and military training. Information regarding distance learning and accredited institutions may be obtained from Distance Education and Training Council, 1601 18th Street, NW, Washington, DC, 20009-2529.

## EMPLOYMENT SERVICES

State public employment offices, which are affiliated with the United States Employment Service of the Department of Labor's Employment and Training Administration, can furnish you with a wealth of information. Several states are now offering a One-Stop Career Center, which will match your aptitudes, background, education, training, and experience against available job openings. These offices can also assist persons in planning a career. Trained personnel or computer data bases give applicants information about the various jobs available, specific job requirements, opportunities for advancement, rate of pay, and other related data. Employment counseling also assists both beginners and experienced workers who wish to change their occupations. The primary purpose of this counseling service is to help job seekers become aware of their actual and potential abilities, their interests, and their personal characteristics.

Often, individuals need additional education and training before they can meet the requirements of certain jobs. The role of the referral-to-training section of the public employment office is to suggest training for job qualification. The services rendered by job placement personnel are just what the title implies. Specifically, they attempt to place workers in jobs for which they are suited, by maintaining close contact with the job market.

### Other Information Sources

Other reliable sources of information and assistance in planning a career include public and school libraries, especially career centers and school libraries, which have books, pamphlets, and magazines containing information about different occupations. Many libraries receive publications from the Bureau of Labor Statistics, the business community, and labor unions. Frequently, officials at the local branches of labor unions can provide occupational information relevant to their particular fields. Employers, personnel officers, and community-based organizations can usually furnish helpful job-related information.

### ASSESSING YOUR OWN ABILITIES

How do you go about assessing your abilities? One suitable, effective method is to carefully analyze all the qualifications considered essential for admission and achievement in the field you choose. Also, you may ask yourself a number of pertinent questions such as: Is my interest in the occupation genuine, or am I merely looking for employment? What is my attitude toward this type of work? Is it too mechanical, technical, or strenuous? Can I take orders and grasp instructions readily? Will my temperament and personality allow me to work effectively as a member of a crew? How do I feel about the classroom training associated with the job? Do I look at the training as an opportunity or as a requirement? Do I understand what will be expected of me, and do I feel that I will be able to measure up to such standards? This is just a sampling of the questions that you can raise and attempt to answer honestly. When you have analyzed your answers, you should be able to determine whether or not you have a real desire to enter the particular field. Many schools have career centers with either self- or computer scoring self-interest inventories. These inventories will match your interests against those in a wide variety of occupations and are very useful in career planning.

## SEEKING THE SERVICES
## OF A GUIDANCE COUNSELOR

As our nation progress technologically, and many industries establish skill standards for both entry and advancement, the choice of an occupation becomes increasingly difficult. Thus, it often is both desirable and necessary to consult with a counselor about your desires for a career in a given industry. Faced with an array of job possibilities, you need to learn about those opportunities that will neither frustrate you nor waste valuable time and effort. Vocational guidance counselors can help you recognize your personal characteristics and capacities and so help you determine whether or not you are suited for the particular type of work in which you are interested. The guidance counselor may also recommend suitable school-to-work programs in which you can learn something about a given occupation by actually working in the field.

On the basis of the collected data from questionnaires, assessment of personal interviews, results from testing, and other pertinent information, the counselor arrives at a meaningful appraisal of your capacities, aptitudes, personal traits, and occupational inclinations. From this information, the counselor is able to recommend a specific field of employment in which you are likely to achieve success.

## CHAPTER 3

# EDUCATION AND TRAINING

Getting the most from preparatory education requires foresight and planning. You must seek the type of training that will allow you to develop a solid base on which to cultivate your talents.

Almost any shop or laboratory course offered in junior or senior high school will be beneficial. Such courses will acquaint you with the correct use of basic hand tools and simple shop equipment. Mathematics and science courses are a must if you are going to enter the electrical field and hope to stay employed. And no one can ever get enough practice on blueprint reading. Cad and cam programs will also help, especially if they deal with mechanics, electrical systems, and electronic drawings. Basic courses should be followed by in-depth courses that allow you to develop fundamental skills and broaden your foundation for future growth and specialization. School-to-work programs will not only help you gain the experience that many employers look for but they will also help you develop a positive mental attitude towards work.

## PREREQUISITE COURSES

An electrician is often called on to perform complex tasks that require knowledge of science and technology. Trainees should prepare themselves by taking courses that further their education in science and mathematics. Courses in these disciplines help individuals grasp subsequent instruction on a more sophisticated level.

Apprentices endeavoring to learn the electrical trade are confronted with many formulas, equations, and computations that necessitate a thorough understanding of mathematics. Courses in mathematics that should prove beneficial to future electrical workers include algebra, trigonometry, and geometry. A study of physics and chemistry also lends itself favorably to the study of electrical theory, since it relates directly to these subjects. To successfully complete an apprenticeship program, a person needs the ability to add, subtract, multiply, and divide whole numbers, decimals, fractions, and mixed numbers. In addition the applicant should know how to convert units, such as square units and English/Metric units, perform direct measurements of objects and distances, and utilize algebra, calculate degrees and angles, and compute area and volume.

Since the work that someone in the electrical/electronic industry performs requires both reading and communications skills, a person will have to obtain essential information contained in safety, installation, and maintenance manuals. You should be able to read complex technical documents, code books, regulations, graphs, charts, and diagrams, and understand and communicate verbal and written instructions and warnings.

Individuals seeking a career in the electrical field will do well to take courses in electronics, air conditioning and refrigeration, welding, estimating materials and supplies, technical report writing, and industrial safety. Many educational institutions, having recognized the need for instruction in these areas, have established specific objectives for each of these courses.

In addition, those desiring to work in this field must be able to meet certain physical requirements, which include hearing warning signals, discriminating between colors, maintain balance and perform activities while on a ladder, use both hands to manipulate small objects, operate equipment that requires two-hands, and be able to reach and stretch to position equipment.

Prerequisite courses should be flexible. They should provide training that is transferable to and usable in different occupations. It is also important that the student pursue these courses in logical sequence.

## INDUSTRIAL ARTS

The underlying purpose of industrial arts in junior and senior high schools is to acquaint students with the industrial environment in which they must live and work. A brochure entitled "Industrial Arts in Education," published by the American Vocational Guidance Association, describes industrial arts as the study of industrial tools, materials, processes, products, and occupations pursued for general education purposes in shops, laboratories, and drafting rooms.

There are, of course, specific training objectives involved in an industrial arts program. Some of the more significant of these goals are: (1) to promote clear analytical reasoning when performing mechanical and constructive tasks; (2) to develop efficient work habits; (3) to cultivate favorable attitudes toward group activities in performing constructive enterprises; (4) to develop an understanding of and an appreciation for industrial materials, processes, and products; (5) to cultivate a wide variety of skills in the use of tools and machines; and (6) to promote safe work habits and proper attitudes toward fellow workers.

Industrial arts education is an element of general education. The specific design of an industrial arts program is not directed toward tangible vocational goals. However, these programs often serve to inspire students who have a desire to create and construct projects through the coordination of their minds and hands. Meaningful industrial arts curriculums can provide an excellent background for young people who intend to pursue a vocational education. Industrial arts also allow an individual to see an industry, such as construction, manufacturing, or service, as a whole, thus providing the person with greater insight into the composition of that industry.

## VOCATION AND INDUSTRIAL EDUCATION

Vocational education is a program of specialized education designed to provide instruction leading to occupational competency at an entry level. The term "vocational education" applies to all forms of education and training that are designed to prepare people for successful employment. Vocational trade, industrial, and technical education includes trade training for industrial occupations.

Vocational-industrial education, like industrial arts, involves educating the whole person. Thus, it complements, rather than competes with, general education. However, unlike industrial arts education, it places major emphasis upon developing salable skills and knowledge. Vocational-industrial education curriculums prepare students for specific occupations, provide industry with a source of trained workers, and provide training and retraining for adults who are unemployed or underemployed.

Because of the upward movement of the general education level in the United States and the advancement of technology and automation, the upgrading of vocational-education was an obligation that had to be fulfilled. Today, many fine trade and technical institutes throughout the nation offer excellent facilities and programs that provide students with the opportunity to pursue occupational training relevant to their needs and desires.

The redirection of the public vocational education program set in motion by the Carl Perkins Vocational Education Act should greatly strengthen the occupational preparation of millions of young people. Specifically, these amendments emphasize vocational training closely related to current job markets.

Federally aided vocational education programs ease the school-to-work transition by providing occupational training oriented to job market needs. Cooperative education is one type of vocational education aimed at easing the school-to-work transition problem. In this type of program, a formal relationship is established between the schools and employers in a given community. Students who participate in this program are allowed to divide their time between school and work. Their school curriculum broadens their general education, while their attendance at job site broadens their experience and create a positive attitude towards work. Another form of education is Tech Prep, or 2-plus-2 education, which divides a person's time between the secondary and postsecondary educational institutions.

The figure below shows the structure of vocational trade and industrial education within the framework of many of our present school systems.

The electrical construction worker, usually referred to as a journeyman inside wireman, does the installing and maintenance of power, lighting, communications, and other systems in commercial, industrial and institutional establishments. This person also does residential work in some geographical area. In other geographical areas, however, residential work is done by a residential wireman. The electrical construction worker is trained through a five-year apprenticeship program that includes training in the following elements:

- Planning and initiating projects
- Establishing temporary power during construction
- Establishing grounding systems

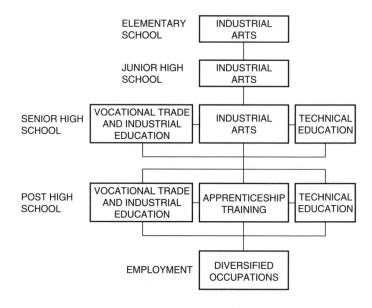

- Installing electric and telecommunication services
- Establishing power distribution systems
- Planning and installing raceway systems
- Installing new wiring and repairing old wiring
- Providing power and control wiring to motors, generators, and other equipment
- Installing and wiring receptacles, lighting systems, and fixtures
- Installing and repairing traffic signals, outdoor lighting, and outdoor power systems
- Installing fire alarm and security systems
- Supervising other electrical construction workers and apprentices

- Installing instrumentation and process control systems and energy management systems
- Erecting and assembling power generation equipment
- Installing, maintaining, and repairing lightning protection systems
- Installing and repairing telephone and data systems

## SAMPLE APPRENTICESHIP SCHEDULES

The following schedules were extracted from a U.S. Department of Labor Employment and Training Administration's trade and industry publication. They describe, in part, the job contents for selected occupations within the electrical industry. These outlines are included in this chapter to acquaint the student with a sampling of the work experiences and training associated with various electrical occupations. These outlines are intended solely as guides.

### Industrial Maintenance Electrician

#### I. ELECTRICAL CONSTRUCTION 2,076 HOURS

Safety instructions
Bend and install conduit
Bend and install other wiring
Run wire and make hook-ups
Install lighting and power circuits
Install power and control equipment
Install and line up motors and generators
Lay out job from blueprint and select material

### II.  GENERAL MAINTENANCE 1,766 HOURS

Safety instructions
Check lights
Electrical equipment
Machinery
Lubrication
Diagnose trouble in lighting and power circuits
Locate cabinets and distribution boxes
Periodically check and repair electrical equipment
Adjust and repair welders

### III.  CRANES AND ELEVATORS 520 HOURS

Safety instructions
Check, repair, and adjust limit switches
Check, repair, and adjust safety devices
Locating and repairing fault electrical equipment

### IV.  ELECTRICAL REPAIR 1,650 HOURS

Safety instructions
Motor repair (A.C. and D.C.)
Test block
Controller repair (A.C. and D.C. starters)
Heating appliances
Building and repairing transformers and welders
Drill repairs

### V. POWERHOUSES: SUBSTATION CONSTRUCTION 850 HOURS

Safety instruction
Heavy cable installation
Lay out and install heavy conduit and duct
Lugging, rubber covered cable splicing, installation of pot
    heads

Connect cable in master and distribution cabinets
Wiring busways and bus
Wiring switchboards and switchgears
Wiring transformers, motors, and generators
Connect instrument transformers, meters, and relays
Check and test circuits

### VI. MAINTENANCE OF POWERHOUSE AND SUBSTATIONS
### 466 HOURS

Safety instructions—servicing of main and auxiliary circuits
Servicing of equipment
Adjusting of the timing of relays
Testing and reconditioning of the transformer and oil switches,
  coolants for transformers and switches
Testing of all circuits

### VII. RELATED INSTRUCTION 672 HOURS

TOTAL 8,000 hours

## Electric Motor Shop Journeyman

### I. SHOP WORK 8,000 HOURS

Shop errands and orientation
Cleaning electrical machinery and auxiliary equipment of vari-
  ous kinds
Stripping old windings, cleaning bar coils and tinning leads
Appliance and fan repairs, locating trouble in fractional H.P.
  motors (A.C.) and repairing
Winding field coils, D.C. armatures, transformers, coils, and
  solenoid coils
Testing field coils and armatures for shorts, opens, and grounds

"Troubleshooting" on various types of equipment in the shop
Winding and forming armature and stator coils, also redesign of same
Commutators and general rebuilding of same
Batteries and battery charging equipment
Starter and control repairs; wind and repair armatures, rotors, and stators; repair brushes, brush holders, and motor leads
General machine shop practices
Assist with trouble shooting in field
Switchboard work
Welding—spot and flame
Taping and insulating coils, also bar windings
Testing various types of electrical apparatus

TOTAL 8,000 hours

## Marine Electrician

### I. WIREWAYS 1,000 HOURS

Layout of raceways—raceway prints
Making and installing hanger—tubes
Cutting threads, bending, and installing conduit
Installing cables
Stripping cables

### II. POWER WIRING 2,000 HOURS

Print reading
Installing motors and controllers
Connecting motors, controller switches, circuit breakers

### III. MARINE LIGHTING 500 HOURS

Layout print reading
Installing fillings and fixtures
Making joints, soldering, taping, and splicing

### IV. INTERIOR AND OTHER COMMUNICATIONS 1,000 HOURS

Bells, batteries, alarm connections, and gyros
Telephones
Engines, order telegraphs; follow-up system
Radio installations
Indicator—prints

### V. INSTALLATION OF GENERATORS AND SWITCHBOARDS 500 HOURS

### VI. YARD MAINTENANCE 2,000 HOURS

Lights, transformers and connections
A-C power, motors, cranes

### VII. OPERATING ELECTRICAL EQUIPMENT 1,000 HOURS

Tests—circuit continuity
Megger, voltage, amperes, measurements

TOTAL 8,000 hours

## Electrical Lineman (Light and Power)

### I. PREREQUISITE EXPERIENCE—
### LOADING AND UNLOADING TOOLS AND SUPPLIES ON TRUCKS; SETTING POLES AND ANCHORS; SENDING TOOLS AND SUPPLIES TO LINEMAN ON POLE; OTHER ASSIGNED WORK ON GROUND

Names and uses of tools and materials
Care of tools and rubber goods
How to tie rope knots and slings
How to make various partial assemblies of materials on the
   ground before installation on the pole

### II. LINE CONSTRUCTION; SETTING, GUYING, AND ANCHORING POLES, AND POLE-TOP WORK

Use and care of climbing tools
Pole climbing skill

Safety precautions, including use of rubber gloves and blankets
First aid and safety methods
Team work, sequence of operations
Public relations and safety
How to arrange for short service interruptions
How to keep job and work order accounts
Pole line construction methods: span lengths and sagging-in, hardware
Tree clearance
Temporary construction schemes to avoid service interruptions
Principles of electricity
Reading electric drawings and sketches, and ability to make a legible sketch or diagram

### III. BUILDING TRANSFORMER STRUCTURES, INSTALLING TRANSFORMERS AND PROTECTIVE EQUIPMENT, AND CONNECTING TO LINES

Construction methods for transformer structures
How to connect and phase out transformers and lines
Knowledge of connections commonly used
How to work on energized lines up to 4,500 volts

### IV. STREET LIGHTING INSTALLATION

Series and multiple circuits
Methods of making installations
TERM    4 years

## Powerhouse Electrician (Light and Power)

### I. ORIENTATION

Learn physical location of electrical equipment
Familiarity with power plant terminology

Knowledge of company organization

Familiarity with symbols and switching diagrams

### II. MAINTENANCE OF ELECTRICAL EQUIPMENT, INCLUDING OIL CIRCUIT BREAKERS, LIGHTING EQUIPMENT, AND STARTERS

Knowledge of routine maintenance required, such as oil changes, cleaning, and inspection

Learn proper oils and greases to be used for specific pieces of equipment

Know how to remove electrical equipment from service and obtain clearances before switching

Proper care of storage batteries

Knowledge of simple lighting circuits and methods of repairing lighting fixtures

### III. CHECK AND MAINTAIN SAFETY EQUIPMENT

Familiarity with various types of fire extinguishers and methods of refilling

Knowledge of first aid and safety methods

Proper use of rubber gloves and blankets

Knowledge of all types of fuses and fuse cartridges in use

Understanding of methods of fighting electrical fires

### IV. PERFORM SHOP AND PLANT DUTIES, AS REQUIRED

Care and use of hand tools

Knowledge of soldering, brazing, and elementary welding (gas and electric types)

Care and use of power tools (lathe, drill, press)

Knowledge and use of insulation materials, and knowledge of checking insulation by use of megger

Use and interpretation of a "growler"

#### V. MAINTAIN RECORDS, AS REQUIRED

Knowledge of purpose or records
Basic accounting for proper distribution of materials

#### VI. INSTALL CONDUIT AND WIRING FOR ANY ELECTRICAL EQUIPMENT

Method of installing conduit
Knowledge and proper use of conduits
Familiarity with the use of proper cable and wire as pertains to size and insulation
Method of handling wire in conduit
Knowledge of conduit sizes and permissible circuits in any size

#### VII. INSTALL AND CONNECT TRANSFORMERS

Methods of installing and connecting transformers
Knowledge of connections and circuits of current, potential, and power transformers
Theory and safety precautions to be observed
Phasing between banks
Checking voltages of various connections

#### VIII. INSTALLATION AND MAINTENANCE OF REGULATORS

Theory of induction and voltage regulators
Method of control
Routine care, methods of installation, and maintenance

#### IX. INSTALLATION AND MAINTENANCE OF ROTATING MACHINERY

Theory, care, and installation methods of A.C., D.C., generators, and motors (includes single-phase, polyphase, synchronous and induction constant and variable speed for A.C. and series, shunt, and compound wound D.C. machines)

Routine care

Methods of checking motor loads

Adjustments and repairs of starters

Methods of checking motor clearances and motor bearings

How to reconnect for different voltages and speeds

How to care for commutators

How to locate and cut out defective motor coils (A.C. and D.C.)

### X. INSTALL METERS, RELAYS, AND CONTROLS

Basic theory of watt-hour, meter, voltmeter, ammeter, and watt-meter

Analysis of control diagrams

Knowledge of relays and remote controls

Adjustments of elevator controls

Knowledge of proper uses of voltmeters and ammeters

Understanding of operation of industrial instruments, such as temperature recorders and remote flow meters

Knowledge of location grounds and short circuits and ability to apply emergency correction

Understanding of when to use "stay put" or other types of push buttons, as well as no-voltage releases, for continuity of service

### XI. PERFORM GENERAL PLANT CONSTRUCTION

Working knowledge of plant power distribution

Knowledge of wiring diagrams

Blueprint reading

Elementary knowledge of electronics and its industrial use

How to plan and choose switch capacities, circuit capacities, controls, and plant installations

TOTAL 8,000 hours

## Sign Electrician

### I. SHOP WIRING

Primary insulation
Secondary insulation
Flashers
Transformers
Time clocks
Switches
Fuse blocks

### II. SHOP ASSEMBLY

Installing tube supports
Bushings
Housings
Mounting tubing

### III. SHOP ERECTING

Prefabrication
Structural steel
Roof signs
Marquee signs
Operations required
Burning
Punching
Welding
Bolting
Coping
Loading

### IV. JOB ERECTING

Layout
Preparation of site for installation

Installation of rigging
Landing
Securing signs
Connection to power supply

### V. SERVICE OR MAINTENANCE

Troubleshooting
Removing defective parts
Replacing defective parts
Repairing, patching, and refinishing
TERM   4 years

## APPRENTICESHIP TRAINING

The individual worker's level of skill will determine the quality of work he or she can do. To develop skills fully, the worker must receive training. One excellent way to obtain it is by becoming an apprentice.

An apprenticeship is the recognized method of learning a skilled trade. In the electrical industry, the vast majority of journeymen have acquired their knowledge and skills through apprenticeship programs. In 1941, the International Brotherhood of Electrical Workers and the National Electrical Contractors Association recognized the need for pertinent apprenticeship training standards for the electrical construction industry, and formed a National Joint Apprenticeship and Training Committee. This committee was assigned the responsibility of formulating training standards, in cooperation with what is now the Bureau of Apprenticeship and Training at the U.S. Department of Labor. Because of the dynamic nature of the electrical industry, it is necessary to review and revise these standards from time to time.

The effectiveness of a training program depends, in large measure, on the training techniques involved. Apprenticeship training combines on-the-job training with related classroom instruction, thus reinforcing the learning process. In current practice, it provides a job for the trainee while providing an opportunity to acquire knowledge and develop skills. Electrical construction apprenticeship training programs for training inside wiremen are based on a five-year curriculum.

The role of the Bureau of Apprenticeship and Training is to assist labor and management in the development, expansion, and improvement of apprenticeship and training programs. The bureau's principal functions are to encourage the establishment of sound apprenticeship and training programs and to provide technical assistance in setting up such programs. The bureau works closely with state apprenticeship agencies, education institutions, and labor and management. Additional information on apprentice programs may be obtained from the local office of your state's employment service.

The importance of apprenticeship training can not be overestimated. Former Secretary of State Ray Marshall, who served as Chairman of the Federal Committee on Apprenticeship, said in an address to the New York State School of Industrial Labor Relations, "... that an expanding and improving apprenticeship system is essential to the welfare of workers and the economic health of the nation...."

## CHAPTER 4

# OCCUPATIONS IN THE ELECTRICAL FIELD

Electrical workers generally fall within two major industry divisions as defined by the Bureau of Labor Statistics (BLS). These two divisions are: the goods producing sector and the service producing sector. The construction and manufacturing industries are part of the goods producing sector. Transportation, utilities, and other services come under the service producing sector. In 1994, these industries accounted for a total of 113 million jobs; they are projected to encompass nearly 130 million jobs in the year of 2005. These industries provide the bulk of job opportunities for electrical workers and other skilled occupations. Skilled workers have higher earnings, more job security, better chances for promotion, and more opportunities in general than most workers without a skilled occupation or professional background.

In 1994, according to the BLS, there were 14 million precision, production, craft, and repair workers. Employment for this important group of workers is expected to reach 14.9 million by the year of 2005—an increase slightly in excess of 10 percent. In this book, we will limit our discussion to selected occupations within the electrical field. The BLS reported that in 1994 there were 388,000 electricians employed, and that this group of workers could reach 591,000 by the year 2005.

The electrical industry encompasses a multitude of occupations and job classifications, and space does not permit the inclusion of all the occupational titles that are relevant to the electrical industry. For a comprehensive listing of job titles, the reader may want to refer to the *Dictionary of Occupational Titles,* published by the U.S. Department of Labor. It can be found in most public libraries and in career centers in high schools. The occupations selected here were chosen because they are key jobs and mirror a wide cross section of job content and skills.

Various electrical jobs have common elements. A qualified journeyman usually has a degree of job mobility within his or her field. The worker may decide to work in a job classification other than his or her normal one; for example, a construction electrician may seek employment as a maintenance electrician, or a maintenance electrician might look for employment as a marine electrician. Within reasonable limits, there are other interchangeable combinations.

In 1992 and 1993, the U.S. Departments of Labor and Education initiated 22 occupational skill standards projects. The skill standards, which are currently being developed, must be compatible with world-class levels of industry performance. They will be tied to measurable, performance-based outcomes that can be readily assessed and be comparable across industries, similar occupations and states. Since these standards will be addressing world-class benchmarks and high performance work organizations, they will have a large impact on the current structure of jobs in industries that adopt them. Currently, out of the 22 projects, 8 to 10 will impact on the electrical-electronic industry. Additional information concerning these standards may be obtained from the National Skill Standards Board, 1441 L Street, N.W., Suite 9000, Washington, DC, 20005-3512.

## CONSTRUCTION ELECTRICIAN

The duties of an electrician are numerous and varied, and present an exciting challenge to individuals seeking a career in this field. *Inside construction electricians* lay out, assemble, install, and test electrical, electronic and telecommunications, security, and fire alarm circuits, fixtures, appliances, equipment, and machinery. They must be able to install electrical lighting, heating, cooling, and control systems in various types of structures: residences, commercial and industrial establishments, schools, hospitals, libraries, and other buildings. They also are frequently called upon to make sophisticated installations involving electrical motors, controllers, transformers, switchgear, and other electrical apparatus.

To cope with their tasks, construction electricians must be able to apply the knowledge learned during apprenticeship training. They also must be able to use diverse electrical formulas and computations associated with their work, such as determining the size of electrical service conductors, feeders, and branch circuits resistance of various sizes of wire and devices, number of conductors that can be put into conduit or in a raceway.

In complying with the requirements of the National Electrical Code, OSHA regulations, and with other safety standards, electricians often must make calculations on starter sizes for motors and other electrical devices. They also must install interior wiring circuits, which include the circuits' conduit or raceway system, the wiring or conductors, fixtures and appliances, and protective devices which ensure safe operation of the system.

In addition to wiring buildings and industrial plants, construction electricians install street lighting systems, motorized equipment for bridges, machinery and wiring for power plants and substations, and sophisticated communications, alarm, and security systems. They also carry on a host of other projects that fall within the domain of the building trades. They must be familiar with the National Electrical Code, which sets forth acceptable standards governing electrical installations. Electricians also must be aware of the existing state, county, and municipal regulations pertinent to their work.

Electricians furnish their own hand tools: screwdrivers, pliers, levels, plumb bobs, hammers, pocket knives, hacksaw frames, compass saws, braces and bits, pipe wrenches, and adjustable wrenches. An initial expenditure of $400 will usually provide a construction electrician with an adequate supply of hand tools. The electrical contractor supplies large tools and equipment, including hydraulic benders, power tools, ladders, and such expendable items as hacksaw blades, taps, and twist drills.

Unlike the other occupations that will be discussed in this book, employment for the construction electrician is somewhat seasonal. There usually are more employment opportunities in the summer than during the winter season, especially in parts of the country where the winters tend to be severe.

Employment practices for the construction electrician also differ in another way. Whereas the maintenance electrician, electric motor shop journeyman, marine electrician, utility and power plant electrician, and electric sign serviceperson change employment infrequently, construction electricians often work for a substantial number of contractors during their careers.

Most construction electricians work for electrical contractors. Some work for government agencies or business establishments that do their own electrical construction work, and a few are self-employed.

The work of the construction electrician, like that of other building trades workers, is active and requires a moderate degree of physical strength, which requires lifting and moving items that weigh up to 50 pounds. It also calls for an alert and attentive mind.

Since a large measure of the work is performed indoors, the inside electrician is not subjected to the elements as often as are other building trades workers.

Construction electricians are crafts workers, and like all genuine skilled workers, they derive satisfaction and pleasure from utilizing both their minds and hands in a constructive manner.

## MAINTENANCE ELECTRICIAN

In essence, the major differences between the construction electrician's duties and those performed by a maintenance electrician lie in the application of their individual skills. The maintenance electrician is responsible for maintaining existing electrical systems, equipment, and machinery. The electrician must periodically service, and, when necessary, make repairs that ensure the continuous operation of the facility in which he or she works. Many firms employing maintenance electricians require their personnel to establish a preventive maintenance program. This program sets forth a specific time schedule to be followed in servicing the plant's machinery and equipment. Servicing machinery and such equipment as motors, generators, controllers, switchgear, and other electrical apparatus normally entails the following processes: cleaning, minor adjustments, replacement of worn parts, and any other necessary repairs.

Many companies operate 24 hours a day. Consequently, they must utilize a work force consisting of three shifts, each electric crew working eight hours per day. Normally, most of the maintenance personnel are assigned to the day shift, which performs the major part of the routine maintenance. However, in order to ensure continuous operation, the second and third shifts also include maintenance crews. Maintenance electricians employed on the second and third shifts are responsible for preventing costly production losses and inconvenience. In emergencies, they may advise management whether or not an immediate shutdown of equipment is necessary, the amount of time needed to repair the equipment, and whether continued operation would be hazardous.

The maintenance electrician's daily work includes tasks other than routine maintenance assignments and performance of emergency repairs. The maintenance electrician must be able to plan, assemble, install, and test electrical, electronic, security, fire alarm, and telecommunications circuits and systems, which frequently include complex sensory, monitoring, and other sophisticated control devices. The work may also include the installation of lighting circuits, distribution and load centers, fuse panels, magnetic starters, safety switches, electric motor controllers, rectifiers, electrical machinery and equipment, computers, tele-data and communication equipment, and many other types of electrical apparatus. Thus, the worker must also be able to interpret and work from blueprints and wiring diagrams.

Like the construction electrician, the maintenance electrician must be familiar with the National Electrical Code, OSHA, and any guidelines set forth by other pertinent regulatory agencies which control installation practices for the specific geographical area. Since a maintenance electrician almost always performs duties for one organization, he or she is usually not required to have the same type of certification as the construction electrician. Licensing and certification will be discussed in a subsequent chapter.

Whenever a piece of electrical equipment or machinery fails to operate properly, the maintenance electrician must locate the trouble. The technique employed in locating circuit, equipment, or machinery faults is known as *troubleshooting*. To help diagnose troubles effectively and expediently, the maintenance electrician utilizes a variety of test equipment. The following are examples of some of the instruments and test equipment used fairly often: voltmeters, ammeters, voltage testers, test lamps, oscilloscopes, watt-meters, power factor meters, and instrument transformers, logic probes, and various digital devices.

In addition to the tools owned and used by the construction electrician, the maintenance electrician's tool chest customarily includes box and open-end wrenches, a ball-peen hammer, punches, calipers, and a socket set. The complement of tools may range from $400 to $500.

Employment for most maintenance electricians is rather stable. Often, they are employed by only one firm during their entire working career. A large measure of their work is performed inside, and they are seldom subjected to severe weather conditions. But because of the nature of their work, they must frequently work in close quarters and may be exposed to unpleasant indoor temperatures.

Maintenance electricians enjoy and take pride in their work, and they characteristically display a high level of ingenuity. The hourly pay for maintenance electricians is generally less than that of construction electricians. However, because of their steady employment, maintenance electricians frequently have greater annual earnings.

In recent years several companies have combined the duties of the maintenance electrician with those of other classifications such as mechanic, plumber, heating and air conditioning technician.

## THE ELECTRIC MOTOR SHOP JOURNEYMAN

Motor shop journeymen are highly skilled artisans. The composition of their work demands that they possess a wide variety of mechanical skills in addition to their electrical competence. Many of the tasks they perform require the use of various types of machines. Shop journeymen must know how to skillfully operate such shop machinery as metal lathes, static and dynamic balancing equipment, drill presses, power saws, hydraulic presses, and an assortment of coil-winding equipment. For this reason, it is essential that a portion of their training include general machine shop practices.

Motor shop mechanics are primarily responsible for the repair and service of electrical machinery and equipment. Basically, their work can be divided into two general categories—work performed within the shop, and duties discharged outside the shop. The first applies to all jobs worked on inside the employer's establishment, and the latter covers those tasks executed on the customer's property. Typical shop work includes rebuilding electrical motors, generators, starters, and controllers; rewinding transformers, relays, magnetic brake coils, and other miscellaneous coil windings; and the fabricating of switchboards and panelboards. Often, rebuilding electrical motors and generators entails, in addition to rewinding the major elements, either the replacing or repairing of mechanical parts such as the shaft of the revolving member (armature) or the bearings which support the armature. In order to accomplish these tasks, the shop journeyman must draw upon experience with the use of such shop equipment as the lathe, drill press, hydraulic press, and other associated machines. The lathe can be used to resurface the shaft of the armature and to make replacement bearings, while the hydraulic press is employed to insert newly machined bearings.

Welding, both gas and electric, is a valuable "tool of the trade" to the motor shop journeyman. Welding skills come in hand when there are broken metal parts in need of repair. Also, the process of welding is very useful in the construction of switchboards, panelboards, and other apparatus.

In addition to being mechanically inclined, shop journeymen must be adept at troubleshooting. That is, they must be able to analyze control circuits quickly, diagnose problems in electrical machinery and equipment readily, and make necessary repairs efficiently. A substantial part of the job hinges on the ability to read and understand complex control wiring diagrams and controller and starter wiring schematics.

To assist in the work, the motor repairman utilizes a number of instruments. Commonly used are the voltmeter, ammeter, tachometer, megohmmeter, frequency meter, and wattmeter. These instruments are invaluable to the shop mechanic. They facilitate the process of locating faults in electrical circuits, machines, and equipment. Proper use and knowledge of test equipment is of the utmost importance to those who want to become expert troubleshooters. Hence, there is much similarity between the maintenance electrician's duties in locating and repairing faults in electrical machinery and equipment and those of the shop journeyman which come under the heading of troubleshooting. However, the expertise needed to perform intricate, internal repairs on electrical machinery usually calls for an electric motor specialist —the shop journeyman. Since these persons are also responsible for making installations involving new apparatus, machinery, and equipment they are called on to repair, they must be aware of the electrical codes and regulations pertaining to their area of work.

A large part of the shop mechanic's duties involve rewinding electrical motors, generators, and transformers. Sometimes the apparatus is rewound in place, especially when the equipment is too large or extremely difficult to move. But more often than not, the equipment is removed from its installed position and transported to the repair shop, where it can be repaired much more efficiently. Motor rewinding requires considerable deftness and the ability to concentrate on complex winding arrangements and connections. Proficient armature and motor winders are a select group; their talents are always in demand.

In order to perform the variety of tasks assigned to them, the shop journeymen must own a fully equipped toolbox. Their set of tools costs approximately $525, and the tools used most frequently are the ball-peen hammer, rule, socket wrenches, calipers and micrometers, an assortment of punches and chisels, pocket knife, combination wrenches, pipe pliers, pipe wrenches, hacksaw, hollow-head wrenches, diagonal and side-cutting pliers, screwdrivers, rawhide mallet, and many special winding tools.

There are times when shop journeymen must work on outside jobs. For the most part, however, their duties are performed under shelter. The physical aspects of the work require strength and endurance, as they must often lift heavy machine parts and do a considerable amount of bending.

The shop journeyman's work is interesting and challenging, and demands a high degree of versatility and skill.

## MARINE ELECTRICIAN

Our country's ships and other water borne facilities are kept lighted and electrified by the efforts of the marine electrician. Almost every operation aboard our modern ships requires electrical power for its accomplishment, and on many ships propulsion is provided by electrical motors that, in turn, receive their energy from electrical generators interconnected through intricate wiring systems. Communications and navigation equipment, pumps and auxiliary machinery, steering gear motors, and a host of other electrical devices all depend on electricity for their operation—and each contributes to the overall performance of a vessel. All this apparatus is connected by miles and miles of electrical cables that are also installed and repaired by marine electricians.

The reliability of a ship's electrical system depends, in large measure, upon the quality and type of maintenance program employed. And the implementation of these programs constitutes a substantial amount of marine electricians' work.

Marine electricians must be versatile. In addition to installing electrical, electronic, and computer systems, they must be adept at modifying existing systems. Quite often, because of new engineering designs or changes in manufacturing process, an exact replacement for a specific piece of equipment may not be available. Thus, the marine electrician must be capable of evaluating the task at hand, recommending specific compensating adjustments, and carrying out the requisite duties. Like other workers who service machinery and equipment, the marine electrician must possess a natural ability for mechanics. Locating grounds and other faults in electrical systems is another phase of the work, especially if the employer specializes in ship repair and service. The characteristics of a successful troubleshooter in this field are similar to those required of the maintenance and motor shop electrician.

Because of the nature of their work, marine electricians must be familiar with the various installation techniques unique to their field. They must be aware of the importance of maintaining watertight seals throughout vessels; therefore, it is important that they possess a thorough understanding of the proper and approved methods for installing cables and fittings throughout a ship. Marine electricians must also be familiar with the United States Coast Guard regulations pertinent to their work.

Many marine electricians find employment in the oil industry's offshore drilling operations. These electricians are hired to service and repair the electrical wiring and apparatus aboard offshore rigs. Occasionally, they are required to make new installations or modifications to the existing electrical system.

The marine electrician usually owns a set of tools comparable to those of the maintenance electrician. In addition, however, he or she must own several other special tools used exclusively by marine electricians. Two such special items are cable skinners, used to remove protective armor from cables, and packing tools, used to pack watertight terminal glands.

Marine electrical work requires frequent bending, kneeling, climbing, and standing. The work, at times, requires heavy lifting. Good health and physical strength are essential for success in this field. Marine electricians work both indoors and outdoors, so they must be able to tolerate varying weather conditions. Sometimes they are required to work on deck jobs in freezing weather, and at other times the job must be done in hot engine rooms.

Marine electricians usually have stable employment. Their work is interesting, exciting, and challenging. Ample opportunities for employment in this field should arise as our nation faces the task of replenishing our merchant marine fleet.

## UTILITY AND POWER PLANT OCCUPATIONS

Utility and power plant workers are responsible for the generation, transmission, and distribution of electrical power to our communities. These workers operate the numerous power stations spread across our nation. They provide the personnel needed to install, service, and maintain the vast array of complex wiring systems that deliver electrical energy for our instant use.

Many types of workers are needed to produce, transmit, and distribute electrical power and to maintain and repair the machinery and equipment used by utility companies—so there is a variety of job classifications within these groups. However, this discussion will be limited to those occupations that require knowledge and skills parallel to those of other electrical workers. They include the power plant electrician (maintenance), the lineperson, the troubleshooter, and the cable splicer.

The power plant is the heart of any electric power system. It houses the machinery and equipment used in the conversion of energy from a mechanical form to an electrical. Electricity is generated primarily in steam-powered plants. Steam-generating plants use coal, gas, oil, or nuclear energy for fuel. Approximately 80 percent of the energy resources used in this country are fossil fuels (coal, oil, and natural gas). The most critical fuels in terms of estimated reserves are oil and natural gas. Of the total energy-producing resources consumed, approximately 25 percent are used to generate electricity.

Hydroelectric generation is another major source of electrical power, and nuclear fuel is certainly an important source of energy in the production of electricity. Today, in power plant construction, the trend is toward building larger and more fully-automated power stations. In recent years, due to mergers and the deregulation of the electric utility industry and independent power producers, the industry has undergone considerable downsizing.

### Power Plant and Substation Maintenance Electricians

The power plant maintenance electrician is one of the key persons included in this group. The duties are similar to those of the maintenance electrician. The one difference is that the power plant employee works on equipment that is either directly or indirectly responsible for the generation of electrical energy. A substation and switching station maintenance electrician takes care of the equipment in the field that is used to reduce voltage levels for the more effective use of electrical energy or to switch transmission and distribution lines to provide for a constant source of electrical power.

Power plant and substation electricians use essentially the same tools as those used by other maintenance electricians. While power plant electricians work is generally performed indoors, substation and switching station electricians work is generally outdoors. On a routine basis a substantial portion of the job centers around preventive measures which help ensure against major and costly breakdowns. These electricians must be able to make quick, accurate decisions in carrying out assignments, especially when faced with an unforeseen combination of circumstances calling for immediate action.

These electricians, like other power company personnel, usually work on a shift basis, since supplying electricity to residences, commercial establishments, and industrial plants is an around-the-clock operation. The personal qualities fundamental to success in this occupation are the same as those given for the maintenance electrician.

Once electricity has been generated, it is ready for use instantly. To accommodate and facilitate its usage, vast networks of electrical conductors are employed to convey power to almost every locality in the United States. Transmission and distribution departments of utility companies are responsible for the installation and maintenance of these systems. To discharge their responsibilities effectively, these two departments generally employ a substantial number of workers. Their duties range from driving trucks and operating pole-setting equipment to supervising complex installations. The purpose here is to explore the duties of three principal classifications of workers: the lineperson, the cable splicer, and the troubleshooter.

## Lineperson

These workers make up the largest single occupation in the utility industry. Line construction—the setting, guying, and anchoring of poles and structures, and the stringing of high tension lines—constitutes a large portion of the lineperson's duties. He or she must be familiar with the pole line construction methods and the terminology associated with this type of work. The job also includes building transformer structures, installing transformers and protective devices, and connecting this equipment to the power lines. Consequently, the worker must possess a thorough knowledge of transformer theory and connections. He or she must also have a working knowledge of the National Electrical Safety Code, which all utilities subscribe to, as well as OSHA requirements.

A lineperson often works from aerial baskets, which are insulated "buckets" attached to a boom that is mounted on the bed of a truck. The aerial basket or bucket facilitates overhead work, since it can usually be readily position near whatever is to be worked on. In some power companies, a lineperson may specialize in particular types of work. Some may be assigned to new construction only, while others may be assigned to crews strictly responsible for repair work. Linepersons may also work from the running boards of helicopters or operate robots from the ground using closed-circuit TV devices.

Frequently, a lineperson is required to perform tasks at critical heights and under hazardous conditions. Also, there are times when he or she must work on energized power lines and equipment under foul weather conditions. When wires, cables, or poles break, it means an emergency call for a line crew, who must often brave the aftermath of a hurricane, storm, or tornado to reestablish the electrical service we depend upon.

Two other types of linepersons working for utilities are the network lineperson and the underground lineperson. Both of these are specialities of the line trade. A network lineperson works in values that feed large commercial and industrial areas usually underground. An underground lineperson usually works on underground residential distribution systems.

### Troubleshooter

A troubleshooter is an experienced lineperson assigned to the special crews that handle emergency calls for service. Since a troubleshooter is basically an experienced lineperson, he or she should possess all the skills and knowledge demanded of a proficient lineperson. In addition, this worker should be very familiar with the power company's transmission and distribution network.

Troubleshooters often work alone when they are attending to service calls. They are usually assigned a service truck that is adequately equipped to handle routine calls. Troubleshooters move from one assignment to another, as ordered by a central office that receives reports of line trouble. Normally, the troubleshooter is radio-dispatched to problem areas as they occur. There are times when the worker merely disconnects or isolates the defective component or circuit from the rest of the system, allowing safe operation of the system—or at least a part of it—until a repair crew can be sent to correct the breakdown. Of course, the specific action taken by the troubleshooter depends largely on the nature and extent of the breakdown.

Other duties of the troubleshooter include the following: (1) maintaining overhead and underground primary lines and equipment, (2) using hot-line tools and rubber protective equipment when working on energized lines and apparatus, (3) testing, refusing, or replacing transformers, fuse cutouts, and lightning arresters to restore or maintain service, (4) clearing shorts or grounds from lines, (5) restoring downed wires and trimming trees that create a hazard to primary or secondary wiring, (6) making temporary repairs to poles, guys, and fixtures, (7) informing supervisors of repairs requiring the services of a line crew, (8) clearing trouble at unattended or automatic substations by switching, refusing, and setting regulators, (9) patrolling and observing the condition of lines, poles, insulators, transformers, cutouts, and other distribution equipment, (10) maintaining street lighting systems, and (11) replacing broken or burned-out street lamps. They also perform routine clerical work, including field and time reports, as well as material requisitions.

## Cable Splicer

Cable splicers install and repair single and multiple conductor insulated cables on utility poles and towers, as well as those buried underground or installed in underground conduits. They make various types of connections between electrical conductors, insulating the conductors and sealing cable joints with a lead sleeve wiped onto the lead sheathing of a cable assembly. The cable splicer must be familiar with the various methods and techniques employed in bonding and grounding lead sheaths of cables for all voltages. He or she must also be aware of the methods of weatherproofing and fireproofing cables. Additionally, cable splicers must be familiar with the various new kinds of splicing kits on the market that are used especially for insulating splices and newer types of cable that do not have a lead sheath.

Journeyman cable splicers' duties also include complicated construction and maintenance work on dead or energized underground distribution conductors and equipment. They install underground services and conduits, breakers and fittings, service boxes, meter boxes, current protective devices and conduits, breakers and fittings, service boxes, potential transformers, and motor and lighting circuits. The cable splicers' job also entails locating and repairing circuit and equipment faults; handling, forming, and bending all types of copper bus work; and mounting heavy equipment. Cable splicers must be familiar with the characteristics of metals and insulating materials used in their field of work. A knowledge of conduit and duct work is also essential. Cable splicers work in manholes, vaults, and on poles when performing their duties, a substantial amount of which pertains to the installation and maintenance of underground lines and overhead cables and the changing of cable systems layouts. They also make all types of polyphase transformer installations, perform phasing and phase rotation tests on polyphase circuits, and, when required, trace and clear grounds and open circuits on cable systems. Occasionally, when work within this job classification is temporarily interrupted or delayed, the cable splicer may be assigned to other work within the occupational group.

The lineperson, troubleshooter, and cable splicer are, at times, required to work from detailed instructions and blueprints. They must be able to exercise judgment and independent thinking in arriving at good, workable decisions. The workers in these classifications must have patience, precision, manual dexterity, and the ability to exercise due care in connection with their work. Initiative and ingenuity are also essential attributes. Often these journeymen supervise other workers assigned to their work crew. Therefore, they must be familiar with the National Electric Safety Code, which is a consensus code covering the construction, repair, and maintenance of overhead and underground electrical conductors and equipment.

The work of lineperson, troubleshooter, or cable splicer is mainly performed outdoors. Occasionally, they are required to work under adverse weather conditions, such as in the aftermath of a hurricane, storm, or tornado, in restoring electrical power to a community. Their duties demand the ability to climb, bend, and work in somewhat awkward positions. They should possess sufficient physical stamina to meet the requirements of their occupation and should also have good visual perception.

Those journeymen performing line work are usually required to furnish climbers, safety belts, body belts, pliers, hammers, wrench and hand connectors, wooden rules, screwdrivers, adjustable end wrenches, 9-inch lineperson's pliers, and wire-skinning knives.

Line construction and maintenance is interesting and adventurous work. It offers a challenging career to those who want to contribute their knowledge and skills to a branch of the electrical industry that is vital to our nation's security and industrial growth.

## ELECTRIC SIGN SERVICEPERSON

Every community, regardless of its size, has an attractive display of electric signs—from elementary signs advertising a neighborhood drugstore to the elaborate array of spectacular signs that line the main section of New York City's Broadway. Since their invention, luminous tube signs and illuminated plastic signs have been accepted as a valuable outdoor advertising medium. These signs advertise the names, products, and services of the hundreds of thousands of factories, stores, restaurants, hotels, theaters, and other business and commercial establishments across the country. They are installed, serviced, and repaired by electric sign service workers who are responsible for ensuring that the equipment they install and maintain operates virtually free of interruption. Hence, the electric sign serviceperson must be capable of discharging duties in an effective and proficient manner.

Electric sign service workers must have the ability to plan on-the-job layouts from detailed prints that illustrate construction design and materials used. They must be familiar with the proper methods of hanging signs. This, in itself, calls for substantial knowledge and skill in the art of sign suspension. Many signs are large and heavy and require special techniques to support them safely in forceful winds. Consequently, the sign journeyman must be aware of the various types of supports and their safe working loads. The sign erector should also have a good understanding of applied physics and should be thoroughly familiar with rigging.

Another area in which the sign serviceperson should have experience is metalworking. The worker must be able to bend sheet metal and plastic into various forms, cut out metal and plastic parts, and be familiar with drilling or punching holes, riveting, and soldering metalwork. It will also be to his or her advantage to be capable of performing operations involving acetylene burning and electric arc welding.

Service calls make up a large portion of the electric sign serviceperson's job. Whenever an electrical sign fails to operate properly, a worker is dispatched to remedy the failure. He or she must be able to diagnose the cause of the trouble expeditiously and to follow up with corrective action. To do this, the worker must have a comprehensive understanding of electric circuitry, as related specifically to electric signs, of transformer theory and application, of gases and their uses, and of mechanics, especially those pertaining to gears, drives, bearings, and other mechanical parts of revolving signs. Electric sign repair workers replace defective electrical wiring, transformers, flashers, time clocks, sockets, tubes and lamps, electric motors, and other mechanical devices and parts.

Electric sign journeymen must be acquainted with the National Electrical Safety Code, particularly those sections relevant to the construction and maintenance of electrical signs. They should also be aware of local building code requirements that cover construction, erection, and maintenance of signs and outdoor display structures with respect to safety, size, and attachment or anchorage.

Another major portion of the electric sign serviceperson's duties comes under the category of preventive maintenance. Many firms purchase service contracts at the time they buy an advertising display; others purchase a sign maintenance contract after the initial purchase. In each case, the sign serviceperson performs periodic routine maintenance on the equipment covered by these maintenance contracts.

Sometimes sign service workers function as customer service personnel. That is, they occasionally suggest to customers ways to increase the attractiveness and visibility of advertising displays. For example, they may recommend changing the color scheme, attaching flashers, or raising the height of the display. Occasionally the worker is also involved in selling service contracts to customers.

Many sign journeymen invest up to $3500 in tools. Their tool-boxes contain basically the same hand tools as those used by the inside construction electricians. The test equipment used is normally furnished by the employer.

Electric sign journeymen must feel comfortable when working at various elevations. Often they must work from scaffolding, and they are, at times, required to use safety belts and lines. Because this type of work is performed almost exclusively in exterior locations, persons aspiring to careers in this field should enjoy working outdoors. They should also possess agility and the ability to climb well and be able to adjust readily to working at different heights and in various positions. The physical aspects of the job require medium strength, an excellent sense of balance, and general good health.

The electric sign industry has good growth potential and should provide ample employment opportunities for those seeking careers.

## CHAPTER 5

# OPPORTUNITIES FOR ADVANCEMENT

Apprenticeship training provides an optimum training system for learning a skilled trade. It gives the trainee an opportunity to progress in an orderly manner from the entrance level to journeyman status. Once the trainee meets the criteria established for entrance into an occupation with apprenticeships, the progression is fairly well planned and assured—*if* the apprentice satisfies the requirements specified for advancement through the various levels of competency.

## APPRENTICESHIP PROGRAMS

Apprenticeship training involves on-the-job training, supplemented by related classroom instruction. Apprenticeship programs range from two to ten years in length. Programs in the electrical industry usually range from two to five years. As previously mentioned, apprenticeship programs in the electrical construction industry, with the exception of residential construction, are five years in length. The five years are usually divided into ten equal intervals, each six months long. Normally, apprentices receive a set percentage of the journeyman's wage rate for each six months of training they complete. The following is an example of an apprentice wage schedule typical of those found in collective bargaining agreements.

First six months—35 percent of journeyman's wage rate
Second six months—40 percent of journeyman's wage rate
Third six months—45 percent of journeyman's wage rate
Fourth six months—50 percent of journeyman's wage rate
Fifth six months—55 percent of journeyman's wage rate
Sixth six months—60 percent of journeyman's wage rate
Seventh six months—65 percent of journeyman's wage rate
Eighth six months—70 percent of journeyman's wage rate
Ninth six months—75 percent of journeyman's wage rate
Tenth six months—85 percent of journeyman's wage rate

Most apprenticeship programs are developed along functional lines, not only to promote the growth of the apprentice's capabilities, but also to meet the ever-changing needs of our economy. Without this kind of organization, it is easy to overlook a training program's primary objectives. Some of the main goals for the development of well-rounded journeymen are pride in craftsmanship, initiative, and ingenuity. These goals must be interwoven into the fabric of the training program. To produce the best skilled craft persons, the apprenticeship programs must be timely and meet the demands of the market place. Therefore, such programs are always under revision.

Apprenticeship is a training system based on a written agreement between the apprentice and the local joint apprenticeship training committee. The apprenticeship training committee usually consists of six members. Three of these members represent the employers, and three represent the union. They are responsible for conducting and supervising the apprenticeship program at the local level. They test, select, and indenture apprentices and register them with the U.S. Department of Labor's Bureau of Apprenticeship and Training or with the state apprenticeship agency. Local joint apprenticeship training committees often employ a training director to assist them in carrying out their duties.

Apprenticeship training committees are responsible for monitoring and evaluating the variety and the quality of the apprentice's performance. They implement job rotation to ensure that the apprentice receives varied experiences with the latest materials, equipment, and construction processes.

Apprenticeship training programs have certain advantages over less formal training programs. This is especially true of programs that are national in scope, since they enhance the quality of the training through the use of standardized training materials.

Completing an apprenticeship gives workers recognized status and dignity. It also provides them with a margin of job security, and often it may increase opportunities for promotion to supervisory positions.

## ON-THE-JOB TRAINING

Aside from completing an apprenticeship training program, trainees may become skilled trades workers by acquiring skills and knowledge through on-the-job experiences. On-the-job training (OJT) is an informal type of training, and so is usually not structured as well as the formal apprenticeship training program. The trainee endeavors to learn the trade by performing a variety of tasks and by assisting other skilled workers in the performance of their duties. Quite often the trainee does not receive any additional off-the-job training or education to supplement his or her work experiences.

There is general agreement that this type of training has drawbacks. It usually does not come under the auspices of a training committee, nor does the trainee enter into a written agreement covering a specified period of training. There is no automatic wage progression and the trainee lacks the protection that he or she would receive in a formal apprenticeship program. The lack of these elements in a training program can present several problems that must be resolved in order for a young worker to become a well-rounded craftsperson. If they do not receive guidance from a policy-making body, their training can lack supervision and direction. Often skilled workers are under such constant pressure to meet production demands that they have little time to devote to the learner. And since on-the-job training is usually unregulated, the trainees do not receive the variety of work assignments essential for the development of a competent and highly skilled worker. Therefore, the time required to advance from the entry level to that of a fully trained worker is usually greater than that required in formal training programs.

If administered in the proper manner, on-the-job training can lead to successful skill development. Adequate job rotation, monetary incentive for advancement, and job security must be included in the overall objectives of an effective training system for skilled workers.

In some situations, on-the-job training offers workers in companies that do not have apprenticeship programs the opportunity to acquire a trade. It also appeals to persons who are not interested in school work, since generally there is no related classroom instruction involved. Although this type of training has played a significant role in the training of skilled workers in some occupations, it has not been a very popular way of attaining journeyman status in the electrical field.

## Adult Education

Many vocational and technical schools and community colleges offer evening courses in addition to the regular, daytime vocational instruction. Evening courses are planned to meet the needs of adult workers attempting to supplement their on-the-job experiences with related theoretical education and training. The students attending these programs have an adult point of view regarding the method of instruction and the contents of the curriculum. Attendance is voluntary and lasts only as long as the students feel they are receiving the instruction they desire. The success of such classes depends mainly on giving students material relevant to their needs in industry. The methods of teaching and the range of material covered should be as flexible as possible, in order to meet the educational aims of the class.

Most of the difficulties that arise in adult education programs can be overcome if the administrator in charge of such education has a good grasp of the programs under his or her control. Above all, the administrator should make certain that the instructional staff possesses the expertise necessary to impart the skills and knowledge that are essential to developing occupational competency.

Attending adult classes can be a fruitful experience for students who are really serious about advancing in their chosen vocation. These classes help bridge the gap between the practical applications of the trade and the theoretical principles that, together, help form a better understanding of the mechanics and science involved. Students enrolled in adult courses have the advantage of drawing upon the experiences of the instructor. They can ask questions and participate in classroom discussions, which can be beneficial to the entire class. Instructors can, at times, shed light on a complex problem that may have arisen on one of the student's jobs. And they often can smooth the transition from a purely theoretical statement to one that incorporates both theory and practicality.

Usually adult certificate programs last for two years, although a few courses may cover a slightly longer period. The contents of these courses normally range from the fundamental aspects of the trade to an introduction to the more intricate theory and science associated with the field. Many of these courses include laboratory work, which allows students to become familiar with the theory and application of electrical test equipment and circuitry, as well as with the principles underlying the experiment. Other adult education courses might consist of a seminar, workshop, or a program as short as a quarter or semester.

The rate at which a student progresses through the course depends largely on individual initiative and aptitude. In some schools, students are allowed to enter the program at intermediate points during the school term, rather than waiting until the subsequent school term. However, once an individual decides to pursue this type of vocational education, he or she should make every possible effort to attend classes regularly and to actively participate in class studies and exercises.

## Distance Education and Training

Distance education and training courses are another way to acquire theoretical training to supplement on-the-job work experience. This type of independent study is widely known. One of the more attractive features is the latitude students have in allocating their spare time for study. However, this relatively unstructured method of study can present serious problems for students who lack initiative and determination. People who enroll in distance education and training courses must discipline themselves to use their spare time prudently. And they should adhere to a study pattern that will allow them to accomplish their vocational objectives within the recommended period of time.

One very desirable feature of this method of study is that students learn by doing. That is, they rely primarily on teaching themselves. This educational technique often helps to instill greater self-confidence. Usually, the course materials are designed to stimulate the student and bring about the desired response. The students, within limits, proceed with their work in accordance with their own abilities, so their study assignments have a fair degree of flexibility. They may choose to spend more time on the subjects that give them the most difficulty and less time on the ones they are able to grasp readily.

Such courses allows students a reasonable amount of freedom in selecting courses that pertain to their major area of interest. Hence, students can subscribe to those courses that serve their immediate needs.

Most programs offer continuous levels of instruction, from basic through advanced courses. And, normally, it is up to the students to decide just how far they want to go in the program, once they have completed their basic block of instruction. For example, the following courses could prove valuable to persons desiring careers in the electrical field: mathematics, drafting, blueprint reading, electricity, electronics, industrial electronics, construction, building maintenance, refrigeration and air conditioning, solar heating, and welding.

Additional information on this type of education can be obtained from the Distance Education and Training Council, 1601 18th Street N.W., Washington, DC, 20009-2529.

## SKILL IMPROVEMENT PROGRAMS

Skill improvement programs usually are geared to advanced training or training in a specialized area above the apprentice level, while the training programs previous described normally apply to training below the journeyman status. Journeyman training programs provide workers with an excellent way of obtaining instruction on current installation practices and techniques, as well as furthering their knowledge of the theory associated with their fields of study.

The technical knowledge gained through attending skill improvement courses often qualifies the participating workers for key and supervisory positions in their field. By concentrating on instruction in a particular phase of their occupation, the journeymen can develop expertise in that area. For example, they may decide that they would like to become proficient in fiber optic systems. To accomplish this aim, they would focus their attention specifically on that type of training.

Skill improvement courses are usually offered by vocational-technical institutes. Quite often, these courses are sponsored by labor unions. Sometimes they are backed by management; occasionally they are conducted on a joint basis. Many skill improvement courses consist of the units of instruction necessary to make the student competent. Upon completion of the various units of instruction, the journeyman is awarded a certificate that signifies that he or she has satisfied the requirements for that particular course.

The main objective of skill improvement programs is to develop additional skills and technical competence. Also of prime consideration is inspiring the craftsperson to continue the quest of knowledge of the trade and to strive to produce quality work.

Skill improvement courses also increase the journeyman's opportunities for promotion to a supervisory-level job. Many supervisors and others in high administrative positions in industry have come from the ranks of crafts workers.

### Military Training

The military should not be overlooked as a source of training for an occupation. To make the service more attractive, many of the branches have instituted apprenticeship training programs. These are registered with the Bureau of Apprenticeship and Training, U.S. Department of Labor.

The military offers both instruction and training in many studies that are not taught in regular trade and technical schools or community colleges. For example, the nuclear navy has been a chief source of applicants for those electric utilities that have nuclear generation. Instrument and watch repair are two other specialities that may be learned in the navy. Aircraft mechanics may be learned in both navy and air force, and electronics may be learned in other branches of the military.

## CHAPTER 6

# SPECIALIZING IN RELATED FIELDS

Skilled workers often find it beneficial to develop specialties in areas related to their occupation. There are times when skilled workers can increase their earning power or their employability by having one or more specialties. For example, a journeyman electrician may decide to become a motor control specialist, a cable splicer, a welder, or an estimator, while a journeyman lineperson may choose to embark on a career as an troubleshooter or a supervisor. The maintenance electrician may find it advantageous to focus his or her talents on the field of instrumentation, electronics, computers, or photonics.

In addition to earnings and employability, interest and aptitude usually play a significant role in determining one's desires to enter into another field. The general working conditions associated with certain specialties often attract individuals into those areas of employment. For example, the duties of someone doing instrumentation may be considerably more challenging and less physically demanding for some workers.

There are several ways to approach specialization. A person might approach specialization directly, learning the occupation much the same way that a skilled worker is trained. However, the best way is to construct a base on which to build. By doing this, a person is capable of working in two or more occupations. This base is essential if the specialist is to impart technical knowledge to other workers.

Persons who want to specialize should try to prepare themselves by taking additional classes and reading the relevant literature about this new field. They should be ready to accept whatever challenges lie within the scope of their abilities, as these challenges will provide them with the opportunity to apply their additional skills and give them added confidence.

## MOTOR CONTROL SPECIALIST

Motor control specialists service and repair motor control equipment and the devices used for starting, accelerating, speed-regulating, braking, and stopping electrical machinery. In earlier years, this type of equipment was serviced by skilled general electricians. However, increased use of electronics, computers, and photonics resulted in the need for specialists who possess expertise in the art of installing, servicing, and repairing intricate control systems. The specialist must not only be familiar with electrical theory and its application, but must also possess knowledge and experience in such fields as electronics, pneumatics, hydraulics, computers, fiber optics, instrumentation, and mechanics.

Today's complex industrial production techniques, which often involve fully automated processes, require knowledgeable personnel capable of minimizing and rectifying control problems, thus reducing costly stoppages of production, waste, and returned products.

Motor control specialists utilize a variety of test equipment when performing their jobs. This equipment includes voltmeters, ammeters, wattmeters, oscilloscopes, signal generators, capacitor checkers, resistance bridges, tachometers, frequency meters, logic probes, and computers. The test equipment used by control specialists often enables them to locate faults that would be virtually impossible to pinpoint without such elaborate test instruments. Of course, they must also rely heavily on sight, smell, touch, and logic. Senses are very valuable when performing any trouble-shooting procedures.

And such specialists *must* be expert troubleshooters. Therefore, they should be thoroughly familiar with all aspects of their field. Some of the devices and apparatus they are responsible for ser-vicing include magnetic amplifiers, saturable reactors, magnetic clutches, amplidynes, Regulex and Rototrol rotary amplifiers, magnetic blowouts, linear and analog amplifiers, microprocessors, computer modules, and a host of other devices used in the control of electrical equipment and machinery. They must also be acquainted with the circuitry in which these devices are utilized, and be capable of reading and understanding schematics, one- and three-line diagrams, and other types of drawings and flow charts used by the electrical and electronic industries.

A major portion of a motor control specialists' work is per-formed indoors. Consequently, they usually work under favor-able environmental conditions. Although some control specialists choose to own special tools (such as nut drivers, small socket sets, bristle brushes, allen wrenches, burnishing tools, contact benders and special screwdrivers) their tools are basically a cross between those used by the inside construction electrician and an electronic technician.

Motor control specialists' employment is fairly stable, since they are often key employees and thus enjoy a degree of job secu-rity. Because they possess knowledge and skills beyond those required for journeyman status, they frequently receive additional compensation in the form of premium pay. This may range from 50¢ to more than $2 an hour above the regular journeyman rate.

The duties of such a specialist are often combined with that of an instrumentation technician. The resulting occupation is called "control and instrumentation technician." A person in this occupation is widely employed by utilities as well as pulp and paper companies.

## ELECTRICAL SERVICEPERSON

The duties of the electrical serviceperson vary from the simple task of replacing blown fuses to the more complex functions of troubleshooting and repairing sophisticated electrical machinery and equipment. Most service workers are assigned trucks that are equipped to handle service calls. Some service workers are radio-dispatched, while others receive their assignments on a daily basis. Like the control specialist, the serviceperson must have a strong background. It is essential, therefore, that these specialists be drawn from the ranks of highly skilled workers. Inside construction journeymen, who are knowledgable of electronic devices, and electronic technician, who know electrical devices as well as the National Electrical Code, that enjoy the challenge of diagnosing circuit and machinery faults usually perform well as service workers. This specialty is also attractive to other journeymen in the electrical field, especially those versed in the art of troubleshooting.

Service workers usually work alone. However, there are times when another mechanic or apprentice will accompany and assist them in handling service calls that require additional workers. The time spent on assigned jobs depends on how difficult it is to resolve the problems. Normally, service workers carry a variety of test equipment in their truck. They also rely heavily on their senses as invaluable troubleshooting aids.

A major part of servicepersons' work is done indoors. Their work is fairly clean and interesting and requires only moderate physical strength. Because of their special skills, they usually enjoy stable employment, and occasionally they have the opportunity to supplement their earnings by working overtime.

Service workers are sometimes permitted to use the company's service truck as their means of transportation to and from work. Their toolboxes contain essentially the same tools used by inside construction journeymen. However, many service workers add special hand tools to their kits to help them make repairs that require specific tools.

Electrical service workers are exposed to a variety of work assignments, each of which offers a slightly different challenge, so they are seldom bored. Also, they frequently derive pleasure from the fact that they have been able to resolve difficult electrical problems that have led to inconvenience or costly shutdowns.

Since their daily work frequently places additional responsibilities on them, and because they provide their own supervision, service workers generally receive a premium of $1 to $3 an hour above the journeyman wage.

## ELECTRONICS JOURNEYPERSON

Over the past 30 years, electronics has become more tightly woven into the field of electricity. At one time, there was very little need for an electrician to have a knowledge of electronics. This is no longer the case. In fact, most electrical trade curriculums now incorporate electronics, telecommunications, fiber optics, and computer-based training. Usually workers who install, service, and repair electrical as well as electronic equipment enjoy favorable working conditions, steadier employment, and in some cases higher wages and benefits than those who only handle electrical jobs. For these reasons, many electricians and other skilled workers enroll in adult skill improvement classes, while others enter individualized training programs, such as distance learning and community college classes.

The electronics journeyperson's duties are diversified. He or she must be capable of performing a variety of tasks—from the simple soldering of a wire to a terminal to more complex assignments, such as diagnosing troubles in sophisticated electronic equipment. These skilled workers must be adept in the art of troubleshooting, repairing, and maintaining both electrical and electronic equipment. Often, a piece of malfunctioning electronic equipment can disrupt an entire assembly-line operation, and unless it is repaired efficiently and put back in service quickly, costly shutdowns can result. Some of the electronic equipment serviced by electronics journeypersons includes electronic precipitators, induction and dielectric heating apparatus, electronic speed control devices for electric motors, various electronic sensing devices used for determining the level and/or temperature of a material, electronic controls used in conjunction with heating and air conditioning equipment, electronic power converters, computers, and telecommunications systems. These are only a few of the electronic devices and equipment the electronics journeyman might be responsible for installing, servicing, and repairing. Some electronics journeymen further specialize in certain areas, such as telecommunications, computer systems and networks, or entertainment equipment.

In addition to the normal test equipment and instruments journeyperson electricians use, electronics journeypersons must be familiar with such test equipment as oscilloscopes, electronic voltmeters, various types of signal generators, power supplies, electronic switching devices, frequency meters and wave analyzers, transistor and IC testers, logic probes, and capacitor testers. Expertise in the use of this test equipment allows a highly skilled worker to pinpoint troubles swiftly in equipment that, without its use, might not be readily detected. The electronics journeyperson's tool kit usually consists of a variety of small hand tools. The toolbox might also contain a number of special-purpose tools designed to perform specific functions.

Electronics journeypersons primarily work indoors. On occasion, however, they will work outside on antennas and pole-mounted electronic devices, such as bridge amplifiers and spitters. While their work does not require great physical strength, it does require a great deal of finger dexterity to work with small parts. The work also requires good eye-hand coordination. These workers must be alert and should have the attributes of a good troubleshooter. Like the control specialist, they customarily receive a premium rate of pay.

## INSTRUMENTATION TECHNICIAN

Another specialty within the scope of the journeyperson electrician's work is the installation, service, and repair of intricate instrumentation and control systems. These systems are employed to monitor, measure, record, and regulate production operations automatically. This field is becoming increasingly attractive to all skilled electricians, simply because he or she receives more exposure to this type of work on today's industrial jobs. As our technology continues to expand, even more opportunities will become available for the highly skilled electrician who possesses sufficient technical competence in this field. These workers install, service, and repair instruments and control devices used to process chemicals; refine oil; control the movements of ships, aircraft, and spaceships; generate electricity; and manufacture steel and a host of other products.

Quite often, sophisticated sensory systems employ a combination of devices that are sensitive to more than one type of energy source. Obviously, industrial instrumentation specialists must be familiar with the disciplines of electricity, electronics, pneumatics, hydraulics, and photonics systems if they intend to successfully meet the challenges offered by this field.

In addition to having a good knowledge of mechanics and electrical and electronic devices (including computers), the instrumentation specialist should also have a good grasp of basic computer programming. Instrumentation specialists are frequently required to work with small components in confined spaces, where they need to use both hands. Many of the minute adjustments and close calibrations they must make require a very precise touch. They should also have the ability to interpret, read, and work from complex circuit drawings.

A significant part of instrument mechanics' work is performed under shelter. However, there are times when the job calls for servicing equipment that is outdoors and in remote locations. Nevertheless, they have stable employment and good earnings.

Like the motor control specialist and the electronics journeyperson, instrumentation technicians utilize a wide variety of test instruments and equipment in troubleshooting, and most of their tools are quite similar to those used in these professions.

Good eye sight, acute hearing, and a keen sense of smell are essential physical characteristics for instrumentation technicians. Usually, their work is not strenuous and does not demand great physical strength. However, it does require good finger dexterity and hand-eye coordination.

Instrumentation technology is intimately woven into the basic fabric of our nation's technological growth. Thus, as our industrial technology expands, so does this branch of the electrical industry. Opportunities for those seeking a career in this specialty are virtually unlimited. Nearly all of our industrial plants utilize systems incorporating process controls, which are serviced by the instrumentation technician.

## TELECOMMUNICATIONS TECHNICIAN

The successful transmission of intelligence over telegraph lines using electrical impulses coded to represent letters and numbers was first demonstrated by Samuel Morse in 1835. Out of this humble beginning grew today's massive telecommunications industry, which encompasses such systems as telephone, radiotelephone, computers, microwave, television, satellite, cellular, cable, and more. In addition, there are in use such specialized systems as radar, sonar, telemetering, and loran that depend on the transmission and reception of electromagnetic waves. The installation, service, and repair of this equipment and its associated circuitry is the responsibility of telecommunications technicians. Some of their other duties include installing, maintaining, and repairing mobile and fixed station radio transmitters and receivers; performing necessary adjustments on microwave equipment; servicing antenna and wave guide systems; analyzing and developing solutions for the various problems that occur in electronic equipment; responding to trouble calls on communications lines; and occasionally climbing poles to make temporary or permanent repairs on lines.

Telecommunications technicians must be adept in the use of test instruments and equipment. They often use circuit analyzers; audio frequency analyzers; multimeters; oscilloscopes; oscillographs; logic probes, and associated transistor and IC testers. Telecommunications technicians identify troubles in the systems and determine signal losses, gains, and levels. They also locate short circuits and defective circuit components, such as defective semiconductors, integrated circuits, transformers, resistors, condensers, and other electronic components. They must be able to read and draw electrical and electronic circuit diagrams. Generally, they use the same hand tools as the electronics journeyperson.

Communications mechanics must have good judgment and ingenuity, and they must be able to be good at customer relations. In addition, they must be able to plan, lay out, organize, and supervise, as well as perform nonroutine work.

Employment opportunities in this branch of the electrical industry are expected to grow rather slowly throughout the end of the decade. Although demand for the services of the telecommunications industry will continue to increase, advances in technology are expected to dampen employment growth. Technological changes in telephone communications should be particularly significant, with computers and other sophisticated electronic equipment playing a greater role in performing automatic functions, thereby reducing manual operations. Nevertheless, there will be always be a number of job openings for those capable of mastering the intricacies of this industry.

Telecommunications technicians work in a variety of surroundings. One day they may be working on a tower, installing a wave guide system for a radar unit or a satellite antenna for data communications. The following day they may be making adjustments on a transmitter in confined quarters. Telecommunications work never ceases to challenge even the most talented troubleshooters, nor is it lacking in interest or excitement. Since the industry changes so rapidly, the skill half-life of someone working in this field has dropped from ten years to five years, and is forecasted to drop to two years by the turn of the century.

The work of a telecommunications technician can on occasion demand physical effort, as when handling microwave dishes or installing equipment cabinets. They should be in sound physical condition and should possess good senses of sight, smell, and hearing. Since the multiple conductor cables have many colors, persons doing this work must be able to properly judge colors.

## SALES OR SERVICE REPRESENTATIVE

Sales or service positions provide interesting and rewarding careers for a substantial number of workers in the electrical utility, telecommunications, and manufacturing industries. The skills offered by these workers are essential for the effective operation of these industries. In addition to being technically competent, sales and service representatives must practice diplomacy. Frequently the job responsibilities require either direct personal contact or telephone contact with suppliers and customers.

Some of the specific duties of sales and service representatives are:

- Promoting the sale of residential, commercial, and industrial electric service
- Providing telecommunications service or the products a company might manufacture
- Negotiating and attempting to secure contracts from customers
- Conducting investigations and advising customers about the use of their electrical equipment
- Arranging for the installation of a system
- Handling service complaints

Customer service personnel should be thoroughly familiar with company products and policies, and should be able to speak clearly. It is helpful to have a temperament that reflects courtesy and tact.

## SUPERVISOR

Those who have attained superior knowledge and skills within their area and have demonstrated leadership qualities are often offered supervisory or team coach positions. Such positions in the electrical industry require a particular set of skills and talents. However, most supervisors usually have two characteristics in common—experience in their respective job areas and the ability to work with others.

Effective supervision is predicated on effective leadership. The adroit supervisor or team leader is not only a master of the trade, but a leader as well. Organizational ability, judgment, integrity, initiative, and interest are essential properties of leadership. Supervisory personnel must possess these traits in order to perform their duties successfully.

Supervisors are responsible for making decisions, giving instructions, evaluating the performance of workers, training, coordinating job functions, and carrying out administrative procedures related to the various records they maintain. Most importantly, supervisors must have concern for the safety and well-being of their co-workers. They must also have the ability to cooperate with team members and assist their subordinates. The degree to which they fulfill these obligations depends, in large measure, on how effectively they apply the basic principles of leadership.

Electrical construction supervisors, line crew chiefs, team leaders, and shop supervisors serve as a link between upper management, workers, and labor union representatives. They must fully understand the terms of the labor agreement under which the members of their team work, so that they can perform their duties in a consistent, fair manner. Besides being adept in communicating with their associates, supervisors must be able to converse intelligently with suppliers, other professionals, and the public. They must be both technicians and diplomats.

The job of a supervisor or team leader in the electrical industry requires the ability to plan, implement, and direct the installation, service, and repair of equipment and systems. Supervisors must be able to transform drawings, instructions, and related information into working systems. They should be familiar with local and national codes, and with safety practices applicable to their branch of the industry. Because supervisors or team leaders perform leadership functions as well as working as artisans in their trade, they receive an additional hourly compensation, which may range from 50¢ per hour to more than $3 per hour over the pay scale of the skilled workers they supervise.

## ESTIMATOR

The talents of accomplished estimators in the electrical industry are always marketable at respectable salaries. Every firm in the electrical contracting and utility business must employ someone who is capable of functioning as an estimator—whether he or she is the owner, a supervisor, or a skillfully trained specialist. In general, most companies find it more efficient and more profitable to entrust the estimating of contract costs to a specialist charged solely with that responsibility. Sometimes such estimating is done by engineers, while at other times it is done by highly trained persons such as draftpersons.

Good estimating requires a combination of engineering knowledge, skill, and practical experience—plus a facility with the tools and techniques of estimating. Electrical estimators must be familiar with the various tables, charts, and graphs used consistently in their field of work. The ability to interpret blueprints and other drawings is a basic necessity for this specialty. Other personal characteristics essential to the estimator are an acute eye and prudent judgment. They must be able to anticipate, within reasonable limits, on-site changes that could substantially alter the cost or nature of a job. Estimates should be accurate, and appraisals of construction, installation, replacement and repair costs must be based on facts. The estimator needs to approximate data as closely as possible and present it clearly and methodically. Personal judgments and guesswork are sometimes unavoidable, but they should be limited strictly to those areas where actual values or conditions cannot be predicted.

Estimators work mostly indoors in an office atmosphere. The job is not characterized by strenuous activity. However, there are times when estimators must work at a demanding pace, especially when they are required to estimate contract costs under an accelerated deadline. Electrical estimators normally have the same workweek as other office personnel, but occasionally their hours of work will differ according to the volume of work they are assigned and the time they are allowed to complete their assignments. Successful estimators enjoy steady employment, good working conditions, and respectable salaries. Frequently, they also share in the profits generated by the firm.

## ELECTRICAL INSPECTOR

Electrical inspectors render invaluable services to municipalities, to the electrical construction industry, and to consumers. Because of the hazardous nature of electricity, many precautions must be taken when installing electrical equipment and systems in residences and other structures. By following the safe practices set forth by various building and electrical codes, hazard-free installations are accomplished. It is the electrical inspector's job to see that this condition is fulfilled. Thus, inspectors are entrusted with the responsibility of determining whether or not electrical installations conform with acceptable standards. Inspectors are usually guided by written codes, such as the National Electrical Code or any other rules and regulations governing electrical work that have been adopted by a regulatory body having jurisdiction within a particular community or defined area.

Electrical inspectors must be knowledgeable in both the technical and practical phases of the electrical trade. They must have an intimate knowledge of the National Electrical Code, as well as any other rules and regulations that apply to their work. Inspectors should be able to communicate effectively with electric power company personnel, electrical contractors and their representatives, and the general public. A substantial portion of their duties are administrative in nature, as they are required to submit reports, write letters, and keep records on their field assignments.

Sometimes electrical inspectors are required to conduct inspections in inconvenient or obscure places. They therefore draw heavily on such personal characteristics as good vision, alertness, and agility to aid them in carrying out physical inspections of electrical installations. Inspectors should be mindful of details— what might, on the surface, appear to be a trivial discrepancy could, if overlooked, result in a catastrophe. Thus inspectors must always be conscious of the responsibility they bear.

Most electrical inspectors are employed by municipalities, but some are employed by states. They enjoy steady employment and favorable working conditions, and their salary is often based on a monthly pay schedule, with the medium range between $1,800 to $2,900 per month. Chief electrical inspectors naturally receive higher earnings. Since the job ordinarily requires the use of an automobile, transportation is furnished or an allowance is given for the use of a private vehicle.

In addition to the satisfaction derived from performing their duties, skilled inspectors enjoy the status of being experts in their field. Many electrical inspectors write manuals and books on subjects related to electrical installations and electrical safety matters.

## ELECTRICAL CONTRACTOR

Electrical contracting can be a lucrative and promising field for those who possess entrepreneurial ability as well as a knowledge of the trade. While most of the specialties discussed so far have focused heavily on the acquisition of additional trade and technical skills, the art of electrical contracting relies fundamentally on the unique combination of business management and technical skills. The experience of contractors and their key employees often makes the difference between success and failure in the electrical contracting industry.

Many a contractor started out at the grass roots level, working with tools as an apprentice or journeyperson. In a number of cases, they have also put in some time as estimators, supervisors, or operations managers. Such practical experience often makes it easier when it comes to planning, scheduling, and controlling business operations. But because the electrical contracting industry is becoming more complex, education and business experience are increasingly important in determining the dollar volume of the business in which contractors are likely to be engaged. A survey compiled by the National Electrical Contractors Association (NECA) shows that nearly half of the biggest contractors in the country are college graduates. More than two-thirds of these graduates earned degrees in engineering, with majors in business administration the runner-up. In short, success as a contractor depends on one's education and experience, plus initiative, foresight, judgment, personality, and the ability to communicate.

## ENGINEER

Engineering is another area of employment within the reach of many of today's electrical worker. High standards of apprenticeship training, coupled with the rise in the general education level of the United States population, provide an excellent base and an incentive for one to further his or her scholastic achievement. Also, the availability of vocational-technical schools, scholarship programs, community colleges and four-year schools bring technical and professional education within the grasp of more people.

Engineers and technicians usually concentrate their technical knowledge and skills in specific areas, such as electrical power, electronics, electrical equipment, photonics, data systems, manufacturing, telecommunications, industrial electrical maintenance, or engineering consulting. This specialization has stemmed, in part, from our country's technological growth, as well as from the development of automated equipment and intricate electrical systems.

Those who want to be engineers or technicians should have an aptitude for mathematics, physics, chemistry, engineering drawing, and mechanics. They must be able to translate engineering concepts into realistic programs and have the tenacity to stick with an assignment and carry it through to completion, no matter how demanding or intricate it may become.

## INSTRUCTOR

Many vocational, trade, and industrial schools, technical institutes, and community colleges recruit teachers from the country's industrial communities. In their selection procedure, the recruiting programs take into account professional training, experience, educational background, technical knowledge, professional attitude, and personality. In selecting personnel for instructorships in either trade or technician training, emphasis is usually placed on the individual's professional background rather than on academic accomplishments. So there are many teaching positions available for craftspeople and other skilled technicians who are capable of imparting knowledge to and cultivating skills in others.

Instructors must understand their field thoroughly and know how to teach it to others. Moreover, they should be able to gear their instructional techniques to reach slow and average, as well as superior students. Teaching careers can be fruitful and rewarding, especially for those engineers, technicians, and artisans who enjoy assisting and working with other people. Many states require vocational instructors to have seven or more years of experience at a trade and pass a trade-related competency exam.

## CHAPTER 7

# EMPLOYMENT AND EARNINGS

Traditionally, the earnings of workers in the electrical industry are as good as or better than the earnings of workers in similar industries. For example, electricians are among the highest-paid workers in the construction industry. Compared to the other construction crafts, electricians work more hours. While there has been some growth in wages over the past five years in all branches of the electrical industry, there has been more steady growth in the earnings of utility workers, maintenance electricians, shop journeypersons, marine electricians, and electric sign service workers in particular. The annual earnings of workers in these branches of the electrical industry provides sufficient income to maintain a comfortable standard of living.

## WAGES

Construction electricians, like most of the other classifications discussed in this book, receive hourly wages. Although the terms *wages* and *earnings* are often used interchangeably, they actually differ in precise meaning. Wage rates stipulate a specific amount of money for a stated period of time (normally an hour), whereas earnings refer to the total amount of compensation received and are affected by the number of hours worked, overtime premiums, holiday pay, and other factors.

Wages and earnings depend upon several factors:

1. The type and nature of work being performed
2. Whether or not the worker is covered by a union contract
3. The geographical section of the country where the work is being performed
4. Who the work is performed for

For example, there might be two employers in a given geographical location employing maintenance electricians, with each em-ployer having their own wage rate.

The accompanying table illustrates the average wage rate for several classifications that have been discussed previously.

**Table 1    Average Hourly Wage Rates for Journeymen (March, 1996)**

| Classification | Industry | Wage Rate* |
|---|---|---|
| Construction Electrician | Inside Construction | $30.13 |
| Sign Serviceperson | Electric Sign | $18.51 |
| Lineperson | Private Utility | $24.60 |
| Troubleshooter | Private Utility | $27.00 |
| Maintenance Electrician | Electrical Maintenance | $14.87 |
| Electric Motor Repairer | Electric Motor Shop | $13.20 |
| Marine Electrician | Ship Building & Repair | $14.56 |

*Wage rates established in collective bargaining agreements in California. Rates will vary across U.S. and Canada.

## FRINGE BENEFITS

A typical contract in the electrical industry provides for substantial fringe benefits, and these benefits play an important part in contract negotiations. A fringe benefit is a benefit paid by an employer over and above the regular wages paid to the employee. Such benefits might include vacations, paid holidays, health care, life insurance, dental care, vision care, paid prescriptions, prepaid legal assistance, educational reimbursement, and pensions. As a result of economic growth, electrical workers are currently enjoying longer vacations, more paid holidays, better pension and insurance plans, and additional sick and bereavement leave. Other benefits might include supplemental unemployment compensation and severance pay allowances.

Vacation plans for construction electricians are normally financed on the cents-per-hour-worked basis described in Chapter 1. Vacation schedules that provide for vacations of one week after six months service, two weeks after one year, three weeks after five years, and four weeks after fifteen years of service are common in some branches of the electrical industry.

The number of paid holidays continues to grow. A recent survey indicated that a large percentage of labor contracts now provide for ten or more paid holidays a year. In some cases, instead of allowing an employee to take off a certain holiday, such as the birthday of a famous person, the holiday is recognized by giving the person his or her choice of days off. This is referred to as a "flowing" holiday. With the exception of electrical construction, workers in the electrical industry receive premium pay for holidays worked.

Supplemental unemployment benefits (SUB) and severance pay allowances are gaining in popularity. These two benefits give the worker a measure of security during such crucial periods as layoffs and terminations. A large number of electrical workers are covered by some type of SUB or layoff benefit plan, and many are covered by some form of severance pay plan. SUBs or layoff benefit plans are usually based on the number of years of service that an employee has with his or her employers. Unless other arrangements are made, it is usually given to the employee in a lump sum, to spend as the employee so desires.

## HOURS OF WORK

The workdays and workweeks of electrical workers mainly depend upon which branch of the industry they work in. Construction electricians usually work straight days, while many workers in other branches of the industry work on a shift basis. Straight day work may be defined as the period of work from approximately 8 A.M. to 4:30 P.M., Monday through Friday, with a half-hour lunch period. Persons on shift work usually work a straight seven- to eight-hour day, eating their lunch or supper on the job. Thus a shift worker might start at midnight and get off at 7 A.M.

A substantial number of labor agreements provide for workweeks of fewer than 40 hours; the number of such contracts continues to grow each year as more and more workers express a desire for longer periods of leisure time. The International Brotherhood of Electrical Workers has over 700 collective bargaining agreements with employers in the electrical industry that provide for workweeks of fewer than 40 hours. These contracts are found in such branches of the industry as construction, maintenance, communications, utility, repair and service, and others. In certain geographical areas and certain branches of the industry the ten-hour day is popular. In this situation, instead of working a five-day week, the person only works four days a week.

## LOCATION OF EMPLOYMENT

The location of employment for electrical workers is determined by the branch of the industry in which they are employed.

For some sectors of the trade, the job site remains relatively constant, while in others the place of work changes and is governed by the nature of the building and construction. Both the geography of an area and its natural resources have an influence on the type of employment likely to be found there.

Highly industrialized areas provide substantial employment for maintenance electricians, electric motor repair workers, utility and power plant workers, and construction workers. An electric sign serviceperson can find good employment opportunities in large commercial areas, where businesses set up huge displays in illuminated plastic and metal fixtures to advertise their products and services. Cities that have a lot of entertainment, such as Las Vegas or Atlantic City, also offer good employment opportunities for electrical sign workers. Port cities—those cities having navigable waterways and shipyard facilities—offer employment to marine electricians.

In summary, the type of industry, the available resources, and the natural features of the land influence the types of employment associated with a given locality.

## EMPLOYMENT SECURITY

The degree of employment security one has in the electrical industry depends on several conditions: the state of our nation's economy, regulations and attempts to deregulate utilities, telecommunications and other industries, weather conditions, and other seasonal factors. The electrical industry, like other industries, is affected in some way by each of these elements, although some other branches are influenced more than others by particular conditions. Moreover, while general economic conditions are likely to affect employment in all branches of the electrical industry, some segments will experience less fluctuation than others.

Employment in the construction industry is more affected by seasonal conditions than most of the other branches of the industry. The lack of building and construction work has been a continuous problem for the construction worker and has contributed to the underutilization of construction resources—especially human resources. Fortunately, employment for the inside construction electrician is one of the more stable building trades, since a substantial amount of the work is performed under cover.

Normally, maintenance electricians and electric motor shop repairers work on a steady basis and are seldom required to take time off without pay. Their work is generally unaffected by inclement weather conditions, and even during business lulls their employers are usually able to keep them gainfully employed repairing idle machinery and equipment or servicing the equipment they use on the job.

Seasonal changes have little influence on marine electrician's employment. However, work in this field is governed, to some extent, by the amount of shipbuilding taking place. Moreover, shipbuilding, like other construction work, lacks constancy. Since many marine electricians also seek employment in the repair and service specialty of the marine field, though, a large number of them enjoy reasonably stable employment.

Workers in the electric utility industry can usually find full-time work. While the number of hours they work per week may vary, from a high in the peak summer periods to a low in less power-demanding winter months, they generally enjoy continued employment. Linemen and other utility workers will average workweeks of 40 hours or more. Currently, the utility industry is in a state of employment flux, caused by increasing number of independent power producers, deregulation of that industry, and a more conservative approach to using electric power.

Because a large portion of the electric sign serviceperson's work is performed outdoors, weather conditions play an important role in determining whether or not the serviceperson can work on a job on any particular day. Although overall conditions of employment for electric sign workers are influenced by both general business conditions and the weather, employment in this field is reasonably steady. Neon sign glass blowers often supplement their employment earnings by blowing novelty glass and models for glass bottles used by other industries.

A high level and range of skills are extremely important for staying employed. If a person is laid off by a certain employer, he or she can then seek employment elsewhere if he or she has the skills that other employers in that area are looking for. A person can never have too many marketable skills.

## JOB OPENINGS

By the year 2005, it is projected that our country will have a civilian labor force of workers 16 years old and older of slightly over 147 million. This reflects an increase of 16 million workers over 16 years old since 1994, or an annual growth rate of 2.3 percent. When employment grows, it spurs the need for additional workers to provide more housing, medical care facilities, educational institutions, and other services and goods. In a similar fashion, as our nation's economy grows, it generates employment opportunities. Obviously, not all segments of the American economy expand at the same growth rate. For example, the service industries are expected to experience the highest percentage of growth through the 1990s and on into the next century. Maintenance electricians, computer technicians, and electronic pagination systems are all projected growth occupations. Another growth area is in the repair and maintenance of electrical and electronic health care equipment.

The capital expenditures required for modernization and growth of industry and service-producing companies will result in more jobs for construction, maintenance, and highly skilled manufacturing workers. The following table outlines the projected growth for some of the occupations discussed in this book.

**Table 2    Employment Projections**

| Occupation | 1994 | 2005 | Increase |
|---|---|---|---|
| All Construction Trades | 3,616,000 | 3,956,000 | 9% |
| Electricians | 528,000 | 554,000 | 5% |
| Communication Equipment Mechanics Installers and Repairers | 319,000 | 266,000 | –17% |
| Electrical and Electronic Equipment Installers, and Repairers | 554,000 | 555,000 | 0% |
| Electrical Powerline Installers and Repairers | 112,000 | 123,000 | 10% |
| Electronic Repairers (Commercial and Industrial Equipment) | 66,000 | 68,000 | 2% |

Source: U.S. Department of Labor, Bureau of Labor Statistics

Employment in goods-producing industries, of which contract construction is a part, is expected to increase by 9 percent by the year 2005. Despite this employment growth, goods-producing industries will not increase as rapidly as the service-producing industries. The Bureau of Labor Statistics expects employment in contract construction to rise from about 3.6 million to about 3.9 million by 2005.

For crafts workers, as a group, employment is projected to reach a little over 17 million—that is, a 4.4 percent increase by 2005. As more industrial plants, commercial establishments, residential, and public facilities are built, the need for more workers in the construction market is created. Numerous job openings are also created for workers to service and maintain these new facilities.

The need for more workers is only a part of the picture of long-term employer needs. A substantial number of job openings result from retirements, deaths, occupational shifts, upgrading and specialization, and other labor force separations. Thus, the total number of job openings for a particular occupation depends both on industry growth and work force replacements.

Detailed employment growth data for electric motor shop journeypersons and marine electricians are not available, but employment in each of these fields is expected to grow moderately over the next decade. Opportunities for workers in these occupations should increase steadily as the nation's economy expands.

The need for electric generation plant personnel is projected to decrease by 3 percent by the year 2005, with the largest decrease being in power generating and nuclear power plant operators. The latter is projected to decrease by 13 percent by 2005.

## TOMORROW'S JOBS

Through the remainder of the 1990s, the job structure for skilled workers in the electrical industry will change drastically in some areas of the industry, as employers desiring to have high performance work places adapt new standards for employment. The contents of most major job classifications within the electrical industry will experience some change with the emergence of new products, equipment, and machinery. New installation methods and other technological accomplishments will also have an influence on industry employment and occupational requirements.

The increasing use of prefabricated components and modular construction techniques will undoubtedly continue through the 1990s, especially in the areas of pre-assembled industrial equipment and residential construction. The net result will be more hours of work performed in manufacturing plants and fewer hours spent at the job site.

Solving environmental problems, advancing space exploration and travel, alleviating transportation problems, and eliminating the uneconomical use of our natural resources and health care will each have a major role in upgrading our standard of living. Activity in these areas will result in the creation of new jobs and changes in the structure of existing jobs. Because much of the technological progress of our nation is the result of our success in electricity and electronics, these industries will probably grow considerably with new discoveries and inventions.

### The Environment

The fight for clean land, air, and water will require long-range planning and prudent policy decisions by our country's leaders. The success of our endeavors largely depends upon several factors. These include:

- The development of new and better electronic precipitators capable of removing larger quantities of pollutants from the atmosphere
- More elaborate water purification systems that will utilize electrical and electronic components
- More sophisticated electrified sanitary centers to deal with the disposal of solid wastes
- The development of electric cars or other means of travel that do not pollute the environment

Another environmental problem is the depletion of rainforests throughout the world. These rainforests, along with the oceans, furnish us with the oxygen we breathe. Thermal-isolation techniques with infrared cameras will allow for greater environmental monitoring and fiber sensors will be used in the near future to sniff out environmental pollutants. These and other environmental control devices will lead to more job opportunities for those workers with the proper education and training.

## Urban Renewal

During the 1990s, substantial progress should be made toward eliminating many of the problems associated with large city transportation systems. A number of cities have either begun installing, or plan to install, modern transportation facilities. The rapid transportation systems presently in operation are highly dependent on the use of electricity. They utilize electrical energy for propulsion and involve elaborate electrical wiring networks and intricate control systems that incorporate computers and microprocessors. Thus, the technology of electricity and electronics will provide the primary thrust for the development of modern, efficient rapid transit systems. Undoubtedly, this area holds significant potential for alleviating our transportation problems and providing substantial job opportunities. Germany, Japan, and other countries are already using rapid electrical rail systems, which are capable of moving people from city to city in excess of 225 miles per hour.

There is a tremendous need to revitalize many of our nation's cities. Such a program is absolutely necessary not only from a human point of view, but also from an economic perspective. Much of this country's industrial production takes place in large cities that are in dire need of rebuilding. Commercial establishments, as well, are in need of repair and replenishment. Housing is another area that certainly needs to be addressed well past the beginning of the next century.

## New Technologies

Telecommunications, entertainment via satellite, digital audio recordings, photonics, electric cars, digital cameras, cellular phones, fiber optic systems, interactive television, a host of new computer designs, and continued space exploration and travel should continue to provide new employment opportunities. While some electrical workers are presently employed in these sectors, these new products and services will create many new opportunities. Some of the more populated occupations are electricians, electronics journeypersons, instrumentation mechanics, test technicians, and control specialists. However, as new technologies are being developed, jobs that were once done by five or six people will now be performed by one. One of the newest job classifications that will emerge will be a *manufacturing technologist.* This will be a person who operates five to six machines, does his or her own quality control, and is available to work with customers who either have a problem or need to be trained to use the newly developed products.

The development of HDTV (High Definition TV) and widescreen TV with theater-quality sound is sure to bring many innovations to those electronic systems that employ basically the same visual technology the television industry presently uses. The considerable improvements in visual quality that will arrive on the scene are likely to cause a revolution in the entertainment and telecommunications industries. This technology could also result in major improvements in the field of radiology, which should result in many additional job opportunities. Electronic switching enables telecommunications companies to expand their business into a host of new markets.

Computer technology also plays an increasingly important role in all industries. Computers and their associated software programs allow companies to transmit all types of data across the nation and around the world. Computer-aided-design (CAD) and computer-aided-manufacture (CAM) is having a major impact on the manufacturing industry. Computers themselves are now in their seventh generation, with a new generation being developed every 18 months. What once took a room to fill now can fit into one's pocket.

Robotics—the use of computers with machine-tool equipment —allows manufacturers to produce relatively small batches of products using programmable automation. Robots now allow electric utilities to do live-line maintenance in storms, inspect the core of nuclear reactors, and perform a host of other tasks that were not possible five years ago. Since robotic technology incorporates the use of computer systems, sophisticated machine-tool equipment, and intricate wiring and control systems, building construction and dangerous maintenance will create new work opportunities should arise for those who are able to produce, install, service, and repair this equipment. This technology is expected to experience substantial growth in both development and application during the rest of the 1990s.

In summary, the remainder of the 1990s should bring forth the opening of a considerable number of new employment opportunities, in addition to the normal annual job openings that occur in the electrical industry.

## CHAPTER 8

# ORGANIZATIONS IN THE FIELD

Each industry is made up of multiple elements, and the electrical industry is no exception. As a whole, it consists of several organizations that represent the concerns of a particular group within the industry. For example, the electric utility industry consists of three or four associations that represent the interest of private investors, public and governmental interest groups, and those companies that generate power from a nuclear source. The purpose of this chapter is to introduce some of the more important organizations associated with the electrical industry. Each has its own particular objectives and functions, but each contributes to the success and progress of the industry. Because of the large number of organizations connected with the electrical industry, only those that are related closely to the occupations described in this book will be highlighted.

## ORGANIZED LABOR

Organized labor represents 16.4 million wage and salary employees, 19.9 percent of all such workers. In private industry, the fields of transportation and public utilities have the highest proportion of workers who are union members, some 27 percent. This is followed by construction and manufacturing, each with 18 percent. Those workers who join unions do so to advance their economic and social interests. Although the early labor movement in this country had few of the characteristics of present labor unions, it did bring workers together to consider problems of mutual concern and to devise means for their solution. It also enabled workers to realize that unity gave them strength, and often allowed them to resist unfair employer demands, such as wage reductions and unreasonable hours of work. As early as 1778, workers had banded together in an effort to secure better wages, minimum wage rates, shorter hours, apprenticeship regulations for crafts, and the advancement of union labor.

The formation of craft unions did not come easily or without stiff employer opposition and much government interference. Under an old English common law doctrine, unions were prosecuted as "conspiracies in restraint of trade." As courts attempted to apply this conspiracy doctrine, it became a very controversial issue. And even when the courts began to back off from the conspiracy issue, they began to focus their attention and restraints on those tools that unions were able to use to secure better wages and working conditions—the strike and the boycott.

## National Unions

Because of the constant opposition faced by unions, and the downward sweeps in our national economy during the first half of the nineteenth century, many of the original unions survived only for brief periods. However, during the 1850s several national unions were founded. By 1859, the Stonecutters, Hat Finishers, Molders, Machinists, and Locomotive Engineers had emerged as national organizations. As national labor unions grew in size and number, collective bargaining began to gain momentum.

By 1850, the workday, often from sunrise to sunset early in the century, was shortened to 10 hours for most skilled artisans in large cities. Moreover, wages for the cities' skilled crafts workers increased, from $1.25 to $1.50 per day in 1820 to $1.50 to $2.00 or more by 1860.

While bargaining with employers was considered to be the most important objective of the early labor movement, labor unions took on additional function—political activity. They became active in government elections, supporting candidates who were sympathetic to their cause and getting laws passed that were favorable to labor. One major social gain made by unions was the extension of citizenship rights. In the decades before the Civil War, the right to vote was denied to working people who did not own property. The unions fought and gradually won the battle, giving these people the right to vote.

### Post-Civil War

The 15 years following the Civil War was a very formative period for the American labor movement. During two cycles of economic recession and revival, 14 new national unions were formed. Union membership rose to 300,000 by 1872, and then dropped to 50,000 by 1878. Three unsuccessful attempts were made to unite the various craft organizations into national labor federations. This period also marked the rise of the eight-hour day movement, which was highlighted in 1868 when Congress established an eight-hour workday for federal employees. By 1886, total union membership again reached the 300,000 level.

In 1869, the Noble Order of the Knights of Labor was founded. It consisted of a small local union of Philadelphia garment workers. From an estimated membership of 10,000 in 1879, the Knights of Labor grew rapidly until it had a national membership of more than 700,000 by 1886. The Knights of Labor's organization was based upon the direct affiliations of local unions with the national associations.

After the emergence of the American Federation of Labor (AFL), the Knights steadily lost ground, and in 1890 they reported only 100,000 members. The Order continued to lose members and finally ceased to be an influential factor in the labor movement, although it existed until 1917.

## American Federation of Labor

By 1881, the nucleus of a new organization had taken shape. The American Federation of Labor favored forming a national organization consisting of national trade or craft unions. In the three decades following 1890, this new organization consolidated its position as the principal federation of American unions. By 1920, the membership of its affiliates had grown to more than 4 million.

The labor movement was, by this time, functioning as a viable organ under the patronage of the federation and accomplishing some of its goals in terms of economic and social improvements. However, it was not free of opposition and setbacks. Nonetheless, as labor unions approached maturity, they were able to obtain the passage of favorable labor legislation. The enactment of the Clayton Anti-Trust Act (1914) provided that antitrust laws should not be construed to prohibit the existence of labor organizations or to prevent labor unions from carrying out their lawful objects. The Davis-Bacon Act (1931) established the prevailing wage in a community for federal or federally-assisted construction. The Norris-LaGuardia Act (1932) established the jurisdiction for courts in matters affecting employers and employees. And the Wagner Act (1935) established the National Labor Relations Board and the first labor law in this country. All these acts were hailed by labor as milestone accomplishments. Each afforded organized labor the opportunity to establish itself more firmly as the medium through which workers could collectively express their needs and desires.

As the membership of the labor movement grew, new problems began to emerge. An internal struggle developed in the American Federation of Labor (AFL) over the question of whether unions should be organized to include all workers in an industry or whether should they adhere strictly to certain crafts or occupations. In 1934 the AFL realized that new methods would have to be used to organize industrial firms using assembly line setups geared for mass production. As a result, the 1934 AFL convention produced a resolution that directed the executive council to issue charters to national and international unions in the automotive, cement, aluminum, and other similarly structured industries, as deemed necessary by the council. The resolution also stated that the jurisdictional rights of existing trade unions would be recognized. Craft unions would continue to organize those industries where skills and work assignments were distinguishable among crafts. However, the industrial or craft organization issue remained unresolved. The problem eventually led to a division in the labor movement, which lasted until the merger between the American Federation of Labor and the Congress of Industrial Organizations (CIO) in December 1955. The Congress of Industrial Organizations was an outgrowth of the committee that had been formed by a few AFL unions to promote the organization of workers on an industry-wide basis rather than along trade or craft lines.

## Labor Legislation

During the transition period when the United States shifted from an economy geared to war production to one focusing on peace, there was a certain amount of industrial strife and work stoppage. This created an undesirable situation for organized labor. It was faced with revived, and greatly strengthened, opposition to the Wagner Act—the most significant labor law that had been enacted thus far. Senator Robert Taft and Congressman Fred Hartley sponsored a rewriting of the act which led to the passage of the Labor-Management Relations Act in 1947. The act, often called the Taft-Hartley Act, became law on June 23, 1947, despite organized labor's strong objections and a presidential veto. In 1959 the Labor-Management Reporting and Disclosure Act (Landrum-Griffin Act) was enacted. The intent of this act is to eliminate or prevent improper practices on the part of labor organizations, employers, labor relations consultants, or their officers and representatives.

Organized labor continued to expand its ranks through the 1950s and 1960s, reaching a membership in the neighborhood of 20 million. Following that period, the ranks of organized labor grew to about 25 million, but membership declined in the 1980s. While the labor movement is still primarily concerned with negotiating with employers and organizing nonunion workers, it is also concerned with public issues. Unions have always encouraged labor's participation in national affairs and have sought to strengthen labor's influence on national policy. Numerous times each year, representatives of the AFL-CIO present labor's views to congressional committees conducting hearings on legislation of interest to the trade union movement. Staff members of the AFL-CIO Department of Legislation and of affiliated unions frequent Capitol Hill to discuss pending bills with individual representatives and senators. Thus, labor unions have become effective in making their voices heard in the nation's legislative councils, as well as at the bargaining table.

In the 1990s an important objective of organized labor will be the promotion of an economic program that will provide American workers with the opportunity for jobs and the right to a full-employment economy and adequate health care. Organized labor clearly recognizes that the underutilization of human resources is a terrible waste. The economic viability of our economy is highly dependent on a fully employed work force. Our nation's industries are more productive when they are operating at full employment and plant capacity. Our social welfare programs work more effectively when there is little unemployment. The need for full employment is apparent. Congress recognized this need and enacted the Humphrey-Hawkins Bill for the purpose of achieving full employment. Adequate health care, a comfortable living for those that wish to retire, affordable education, and clean air and water are high on the agenda for the labor movement

Organized labor will continue to work for passage of legislation that is beneficial to all workers of our nation. Laws that establish minimum wage levels, protect workers' pension benefits, workplace environments, safety and health, fair trade, and public education are examples of the kinds of legislation organized labor supports. Organized labor has long been the champion for public education and laws that provide for social justice. The American labor movement has always stood for a strong economy and industrial base. Just as in previous decades, organized labor in the next century will continue its efforts to fulfill these objectives.

## INTERNATIONAL BROTHERHOOD
## OF ELECTRICAL WORKERS

The International Brotherhood of Electrical Workers (IBEW) is the union that represents most of the organized workers employed in the various occupations discussed in this book. The IBEW was founded in 1891 in St. Louis, Missouri, by a group of electrical workers who were seeking to improve their working conditions. Out of this nucleus has grown an organization that today encompasses a membership of 750,000, including 1,400 local unions spread across the United States and Canada.

IBEW members are employed in every facet of the electrical industry. Some of the major areas in which they are employed include inside and outside construction, utilities, telecommunications, railroads, electrical manufacturing, recording, radio and television broadcasting, and government facilities. Many members are engaged in the crucial work of keeping Americans and Canadians healthy and their countries strong. Some work to bring you live performances and sports on your radio and television and on nuclear projects. Others do x-ray and other medical work. In addition, thousands of IBEW members are involved in the national space program, handling timing and firing, instrumentation, data acquisition and processing, and telecommunications support services.

The IBEW is structured to give every member a voice in its operations. Hence, members have the opportunity to express their opinions through the officers they choose to represent them at the local union level. In turn, the International Union, an affiliate of the AFL-CIO, is able to register the views of its entire membership through the federation. This results in an effective and influential body capable of accomplishing many of its objectives.

The objectives of the IBEW can be outlined best by restating them as they appear in its constitution:

> To organize all workers in the entire electrical industry, including all those in public utilities and electrical manufacturing, into local unions; to cultivate feelings of friendship among those of our industry; to settle all disputes between employers and employees by arbitration, if possible; to assist each other in sickness or distress; to secure employment; to reduce the hours of daily labor; to secure adequate pay for our work; to seek a higher and higher standard of living; to seek security for the individual; and by legal and proper means to elevate the moral, intellectual, and social conditions of our members, their families, and dependents, in the interest of a higher standard of citizenship.

## NATIONAL ELECTRICAL
## CONTRACTORS ASSOCIATION

The National Electrical Contractors Association (NECA) is a nationwide trade association that represents the electrical contracting industry. The association provides management and technical service to 4200 members, most of whom are small business people. As individuals, they generally cannot afford such a wide range of services themselves, but, by sharing costs with others within the association, member electrical contractors are able to enjoy many benefits at a substantial savings.

Proper management services are very important to the small business operator. Some of the services provided by NECA include representation in labor relations, marketing services, technical training services, business management training, public relations, information services, and a field service that covers the whole range of NECA services. NECA's main purpose is to make this information available to electrical contractors and to motivate them to use it to their advantage (and thus to the advantage of the consumer).

For nearly 90 years, the National Electrical Contractors Association has dedicated itself to a constant effort to improve the quality of electrical service offered to the American public and to industry. NECA and the International Brotherhood of Electrical Workers cosponsor the National Joint Apprenticeship and Training Committee for the Electrical Contracting Industry. This committee sets standards and assists local Joint Apprenticeship and Training Committees, to ensure that, in every locality, an adequate supply of thoroughly trained electrical workers will be available to serve the industry's needs.

In 1920, NECA and the International Brotherhood of Electrical Workers formed a joint committee to deal with the various problems that arise in the field of labor-management relations. This joint committee, consisting of equal employer and union representation, became known as the Council on Industrial Relations (CIR) for the electrical contracting industry. CIR is, in essence, a judicial body rather than a mere arbitration organ. Since its inception, it has functioned in an effective manner, reconciling disputes between the employer and the union, thereby resulting in a virtually strikeless industry.

## NATIONAL ELECTRIC SIGN ASSOCIATION

The National Electric Sign Association (NESA) is a nonprofit corporation that promotes the welfare of the electric sign industry. In 1944, the present National Electric Sign Association—now composed of most of the major electric sign producers, suppliers, and component-part manufacturers—was organized. The association has continued to gain strength as a guiding force and voice for the industry. Groups represented by NESA include custom and national sign companies, sign product manufacturers, sign supply distributors, affiliated state associations, advertisers (sign users), and related organizations.

NESA provides its 1200 members with current industrial data such as economical and statistical analyses pertinent to the electric sign industry. The association has a file on job classifications within the sign industry. It also issues a monthly newsletter and publishes a membership directory as well as other special publications.

## EDISON ELECTRIC INSTITUTE

The Edison Electric Institute (EEI) is an association of investor-owned electric light and power companies in the United States.

The institute's objectives are advancement in the public service of producing, transmitting, and distributing electricity; promotion of scientific research in these fields; ascertaining and making available to members and to the public the data and statistics relating to the electric industry; and aiding operating company members in generating and selling electric energy at the lowest possible price commensurate with safe and adequate service, giving due regard to the interests of consumer, investor, and employee.

The EEI was organized in 1933, and since that time it has been a strong, continuous stimulant to the advancement of the art of making electricity. Through the years, as new devices and techniques are developed, their benefits are made known to all electric companies and are passed on by the companies to customers throughout the country. The EEI is the forum through which this information is exchanged.

Besides the EEI there are two other national association of electric utilities. The American Public Power Association, with its membership of 1750, serves municipally owned electric power companies, public utility districts, state and county-owned electric systems, and some rural electric companies. The National Rural Electric Cooperative Association has a membership of 1000 rural electric cooperatives in 46 states. The NRECA activities include professional conference, training, and consulting services and is the legislative spoke person for its membership.

## INTERNATIONAL ASSOCIATION
## OF ELECTRICAL INSPECTORS

The International Association of Electrical Inspectors (IAEI) is a nonprofit organization cooperating in the formulation and uniform application of standards for the safe installation and use of electricity. The association has a membership of more than 18,500, representing every branch of the electrical and allied industries. Some of its major objectives are as follows: formulation of standards for the safe installation and use of electrical materials, devices, and appliances; promotion of uniform understanding and application of the National Electrical Code, other codes, and any adopted electrical codes in other countries; promotion of uniform administrative ordinances and inspection methods; collection and dissemination of information relative to the safe use of electricity; representation of electrical inspectors; cooperation with other national and international organizations in the further development of the electrical industry; promotion of cooperation among inspectors, inspection departments, the electrical industry, and the public.

## NATIONAL FIRE PROTECTION ASSOCIATION

Since its inception in 1896, the National Fire Protection Association (NFPA) has been a nonprofit, voluntary membership organization. Its membership includes more than 65,000 members worldwide. It is recognized internationally as a clearinghouse for information on fire prevention, fire-fighting procedures, and means of fire protection, and as an authoritative source for fire loss experience.

Membership in NFPA is open to all those interested in promoting the science and improving the methods of fire protection and prevention. Its membership is widely representative of industry, commerce, the fire services, government agencies, the military forces, architects, engineers, the other professions, hospital and school administrators, and others. NFPA's basic objective is to provide humanity with a fire-safe environment, using scientific techniques and education. Its basic function is the preparation of consensus standards and codes relating to fire protection and prevention, with safety the paramount concern.

NFPA has an Electrical Section that provides an opportunity for members interested in electrical safety to become better informed and to contribute to the development of the National Electrical Code and other NFPA electrical standards.

The National Electrical Code (NEC) is sponsored by NFPA under the auspices of the American National Standards Institute (ANSI). A primary function of the National Electrical Code Committee is developing periodic revisions of the NEC and other codes related to electrical fire safety. The National Fire Protection Association has acted as sponsor of the NEC since 1911. The original code document was developed in 1897, as a result of the united efforts of various insurance, electrical, architectural, and allied interests.

The purpose of the National Electrical Code and other codes are the practical safeguarding of persons and of buildings and their contents from hazards arising from the use of electricity for light, heat, power, radio and signaling, and other purposes. The NEC is purely advisory, as far as the NFPA and ANSI are concerned, but is offered for use in law and for regulatory purposes in the interest of life and property protection. Further information can be obtained from: The National Fire Protection Association is located at Batterymarch Park, Quincy, Massachusetts, 02269.

## NATIONAL ELECTRICAL
## MANUFACTURERS ASSOCIATION

The National Electrical Manufacturers Association (NEMA) is the industry spokesperson on standardization matters for its seven product divisions: building equipment, electronics, industrial equipment, insulating materials, lighting equipment, power equipment, and wire and cable.

NEMA is the nation's largest trade organization for manufacturers of electrical products, and includes 600 member companies located throughout the United States. All are domestic firms, varying in size from comparatively small companies to diversified industrial giants.

Because standards are recognized as essential to industrial progress, NEMA devotes a substantial portion of its budget and a major portion of its staff time to engineering and technical programs. As a result, NEMA today occupies a prominent and highly respected position in both domestic and international standardizations.

The association is a strong advocate of voluntary standards as being necessary to provide sound and safe electrical products to all people in all markets. At the same time, it recognizes that "any standards are good only to the degree that they assist in performing the desired job more efficiently, more economically, and to the greater satisfaction of all concerned."

## REGULATORY AND LICENSING AUTHORITIES

Many states, and most of their larger political subdivisions, have adopted regulations and standards governing electrical installations. However, in some instances, these regulations do not apply to certain types of installations. Usually, when either a state or city has an electrical code—normally the National Electrical Code (NEC)—some means of enforcement is also provided.

A large number of states and municipalities attempt to regulate the conformity of electrical work with existing codes by licensing those who are qualified to interpret the pertinent codes. Most states, and a substantial number of cities, require that electrical contractors be licensed, since they are primarily responsible for the electrical work performed by the workers in their employ. A lesser number of states and communities require that journeymen electricians also be licensed. Many states and cities employ electrical inspectors who are given the responsibility of determining whether or not electrical installations meet the requirements of the National Electrical Code and/or other relevant codes. Some communities employ private inspection agencies.

The National Electrical Code has been widely adopted by state law or city ordinance. It is universally recognized throughout the United States and is the basis for practically all the legislation adopted regarding electrical installations in buildings.

# SOURCES OF ADDITIONAL INFORMATION

Edison Electric Institute
  701 Pennsylvania Avenue, N.W.
  Washington, DC 20004

Electrical Apparatus Service Association, Inc.
  1331 Baur Boulevard
  St. Louis, MO 63132

International Brotherhood of Electrical Workers
  1125 15th Street, N.W.
  Washington, DC 20005

National Electrical Contractors Association
  3 Bethesda Metro Center Suite 1100
  Bethesda, MD, 20814

National Joint Apprenticeship and Training Committee
  16201 Trade Zone Avenue
  Suite 105
  Upper Marlboro, MD 20772

U.S. Department of Labor
  Bureau of Apprenticeship and Training
  200 Constitution Avenue, N.W.
  Washington, DC 20210

# BUREAU OF APPRENTICESHIP AND TRAINING OFFICES

## STATE OFFICES

| Location | States served | |
|---|---|---|
| Room 510<br>John F. Kennedy<br>  Federal Building<br>Government Center<br>Boston, MA 02203 | Connecticut<br>Maine<br>Vermont | Massachusetts<br>New Hampshire<br>Rhode Island |
| Room 602<br>201 Varick Street<br>New York, NY 10036 | New Jersey<br>New York | Puerto Rico<br>Virgin Islands |
| Gateway Bldg.<br>3535 Market Street<br>Philadelphia, PA 19104 | Delaware<br>Maryland<br>Virginia | Pennsylvania<br>West Virginia |
| Room 418<br>1371 Peachtree Street, N.E.<br>Atlanta, GA 30309 | Alabama<br>Florida<br>Georgia<br>Kentucky | Mississippi<br>North Carolina<br>South Carolina<br>Tennessee |

| Location | States served | |
|---|---|---|
| Room 758<br>Federal Building<br>230 South Dearborn Street<br>Chicago, IL 60604 | Illinois<br>Indiana<br>Michigan | Minnesota<br>Ohio<br>Wisconsin |
| Room 502<br>Federal Building<br>525 Griffin Street<br>Dallas, TX 75202 | Arkansas<br>Louisiana<br>New Mexico | Oklahoma<br>Texas |
| Room 1100<br>Fed. Office Bldg.<br>911 Walnut Street<br>Kansas City, MO 64106 | Iowa<br>Kansas | Missouri<br>Nebraska |
| Room 476<br>New Custom House<br>721 19th Street<br>Denver, CO 80202 | Colorado<br>Montana<br>Utah | North Dakota<br>South Dakota<br>Wyoming |
| Room 715<br>71 Stevenson St.<br>San Francisco, CA 94105 | Arizona<br>California | Hawaii<br>Nevada |
| Federal Office Building<br>Room 8018<br>909 First Ave.<br>Seattle, WA 98174 | Alaska<br>Idaho | Oregon<br>Washington |

## FEDERAL OFFICES

### Alabama

Suite 102
Berry Building
2017 2nd Avenue, N.
Birmingham, AL 35203

### Alaska

Room C-528
Federal Building and Courthouse, Box 37
701 C Street
Anchorage, AK 99513

### Arizona

Suite 302
3221 North 16th Street
Phoenix, AZ 85016

### Arkansas

Room 3014
Federal Building
700 West Capitol Street
Little Rock, AR 72201

### California

Room 350
211 Main Street
San Francisco, CA 94105

## Colorado

Room 480
U.S. Custom House
721 19th Street
Denver, CO 80202

## Connecticut

Room 367
Federal Building
135 High Street
Hartford, CT 06103

## Delaware

Lock Box 36
Federal Building
844 King Street
Wilmington, DE 19801

## Florida

Room 1049
City Centre Building
227 North Bronought Street
Tallahassee, FL 32301

## Georgia

Room 418
1371 Peachtree Street, N.E.
Atlanta, GA 30367

## Hawaii

Room 5113
300 Ala Moana Boulevard
Honolulu, HI 96850

## Idaho

Room 493
P.O. Box 006
550 West Fort Street
Boise, ID 83724

## Illinois

Room 758
230 S. Dearborn Street
Chicago, IL 60604

## Indiana

Room 414
Federal Building and U.S. Courthouse
46 East Ohio Street
Indianapolis, IN 46204

## Iowa

Room 637
Federal Building
210 Walnut Street
Des Moines, IA 50309

## Kansas

Room 235
Federal Building
444 S.E. Quincy Street
Topeka, KS 66683

## Kentucky

Room 187-J
Federal Building
600 Federal Place
Louisville, KY 40202

## Louisiana

Room 1323
U.S. Postal Building
701 Loyola Street
New Orleans, LA 70113

## Maine

Room 408-D
Federal Building
P.O. Box 917
68 Sewall Street
Augusta, ME 04330

## Maryland

Room 1028
Charles Center–Federal Building
31 Hopkins Plaza
Baltimore, MD 21201

## Massachusetts

Room 1703-B
JFK Federal Building
Government Center
Boston, MA 02203

## Michigan

Room 657
Federal Building
231 W. Lafayette Avenue
Detroit, MI 48226

## Minnesota

Room 134
Federal Building and U.S. Courthouse
316 Robert Street
St. Paul, MN 55101

## Mississippi

Suite 1010
Federal Building
100 West Capitol Street
Jackson, MS 39269

## Missouri

Room 547
210 North Tucker
St. Louis, MO 63101

## Montana

Room 394–Drawer # 10055
Federal Office Building
301 South Park Avenue
Helena, MT 59626-0055

## Nebraska

Room 700
106 South 15th Street
Omaha, NE 81020

## Nevada

Room 311
Post Office Building
P.O. Box 1987
301 East Steward Avenue
Las Vegas, NV 89101

## New Hampshire

Room 311
Federal Building
55 Pleasant Street
Concord, NH 03301

## New Jersey

Room 339
Military Park Building
60 Park Place
Newark, NJ 07102

### New Mexico

Suite 16
320 Central Avenue, S.W.
Albuquerque, NM 87102

### New York

Room 810
Federal Building
North Pearl & Clinton Avenues
Albany, NY 12201

### North Carolina

Room 376
Federal Building
310 New Bern Avenue
Raleigh, NC 27601

### North Dakota

Room 344
New Federal Building
653–2nd Avenue, N.
Fargo, ND 58102

### Ohio

Room 605
200 North High Street
Columbus, OH 43215

## Oklahoma

Suite 202
1500 S. Midwest Blvd.
Midwest City, OK 73110

## Oregon

Room 526
Federal Building
1220 S.W. 3rd Avenue
Portland, OR 97204

## Pennsylvania

Room 773
Federal Building
228 Walnut Street
Harrisburg, PA 17108

## Rhode Island

100 Hartford Avenue
Federal Building
Providence, RI 02909

## South Carolina

Room 838
Strom Thurmond Federal Building
1835 Assembly Street
Columbia, SC 29201

## Tennessee

Suite 101-A
460 Metroplex Drive
Nashville, TN 37211

## Texas

Room 2102
VA Building
2320 LaBranch Street
Houston, TX 77004

## Utah

Room 1051
1745 West 1700 South
Salt Lake City, UT 84104

## Vermont

Suite 103
Burlington Square
96 College Street
Burlington, VT 05401

## Virginia

Room 10-020
400 North 8th Street
Richmond, VA 23240

## Washington

Room B-104
Federal Office Building
909 First Avenue
Seattle, WA 98174

## West Virginia

Room 310
550 Eagan Street
Charleston, WV 25301

## Wisconsin

Room 303
Federal Center
212 East Washington Avenue
Madison, WI 53703

## Wyoming

Room 8017
J. C. O'Mahoney Federal Center
P.O. Box 1126
2120 Capitol Avenue
Cheyenne, WY 82001

# HELPFUL PUBLICATIONS

### NATIONAL APPRENTICESHIP PROGRAM

U.S. Department of Labor/Employment and Training
  Administration
Bureau of Apprenticeship Training
200 Constitution Avenue, N.W.
Washington, DC 20210

### APPRENTICESHIP INFORMATION

U.S. Department of Labor/Employment and Training
  Administration
200 Constitution Avenue, N.W.
Washington, DC 20210

### APPRENTICESHIP—PAST AND PRESENT

U.S. Department of Labor/Employment and Training
  Administration
Bureau of Apprenticeship and Training
200 Constitution Avenue, N.W.
Washington, DC 20210

### ELECTRICAL APPRENTICESHIP

National Joint Apprenticeship and Training Committee
  16201 Trade Zone Avenue, Suite 105
  Upper Marlboro, MD 20772

**DIRECTORY OF POST SECONDARY SCHOOLS:**
**WITH OCCUPATIONAL PROGRAMS**

U.S. Department of Education
 National Center for Education Statistics
 Washington, DC 20402

**OCCUPATIONAL OUTLOOK HANDBOOK: BULLETIN 2450**

U.S. Department of Labor
Bureau of Labor Statistics

(This handbook can usually be found in the office of high school guidance counselors and in public libraries.)

**U.S. INDUSTRIAL OUTLOOK**

International Trade Administration
 U.S. Department of Commerce, Superintendent of Documents
 P.O. Box 371954
 Pittsburg, PA, 15250-7954

# GLOSSARY

**Agreement, collective bargaining.** A written agreement (contract) arrived at as the result of negotiation between an employer or a group of employers and a union. An agreement sets the conditions of employment—wages, hours, fringe benefits—and the procedure to be used in settling disputes that may arise during the term of the contract. The term of a contract may be for one, two, or three years.

**Apprentice.** Usually a young person who enters into agreement to learn a skilled trade and to achieve a journeyman status through supervised training and experience, usually for a specified period of time. Practical training is supplemented by related technical off-the-job instruction.

**Building trades.** The skilled trades in the building industry. These include: electricians, carpenters, plumbers, painters, plasterers, bricklayers, and stonemasons. These crafts generally have well-developed apprenticeship programs.

**Collective bargaining.** A method of determining conditions of employment by negotiation between representatives of the employer and union representatives of the employees. The results of collective bargaining are set forth in a collective bargaining agreement.

**Compensation.** A concept sometimes used to encompass the entire range of wages and benefits, both current and deferred, which workers receive out of their employment.

**Craftsperson.** An artisan; a person skilled in the mechanics of his or her craft.

**Earnings.** The total amount of remuneration received by a worker for a given period as compensation for work performed or services rendered, including incentive pay, premium pay for overtime, shift differentials, bonuses, and commissions.

**Fringe benefits.** Nonwage benefits and payments received by or credited to workers in addition to wages. Examples are: supplemental unemployment benefits, pensions, travel pay, vacation and holiday pay, and health insurance.

**Journeyman.** A skilled worker, generally having mastered his or her trade by serving an apprenticeship.

**Median wage.** The wage rate that occupies the middle position in an array of wage rates. To illustrate the distinction between *Median* and *Average*, suppose five persons have wage rates respectively of $8, $9, $10, $13, and $15 an hour. The average wage rate is $11 per hour, while the median wage rate is $10 per hour.

**Overtime.** Work performed in excess of basic workday or workweek, as defined by law, collective bargaining agreement, or company policy. Often applied to work performed on Saturdays, Sundays, and holidays at premium rates of pay.

**Service calls.** A term generally applied to the work performed by a serviceperson or craftsperson at the customer's premises.

**Shift work.** The term applied to the daily working schedule of a firm or its employees. Day shift—usually the daylight hours; evening shift—work schedule ending at or near midnight; night (graveyard) shift—work schedule starting at or near midnight.

**Troubleshoot.** The term applied to locating and/or diagnosing faults in electrical circuits appliances, machinery, and equipment.

# STATE APPRENTICESHIP AGENCIES

## Arizona

Apprenticeship Services
Department of Economic Security
438 West Adams Street
Phoenix, AZ 85003

## California

Division of Apprenticeship Standards
455 Golden Gate Avenue
Room 1169
San Francisco, CA 94101

## Connecticut

Apprenticeship Training Division
Labor Department
200 Folly Brook Boulevard
Wethersfield, CT 06109

### Delaware

Apprenticeship and Training Council
Department of Labor
Division of Employment and Training
P.O. Box 9499
Newark, DE, 19714-9499

### District of Columbia

Director of Apprenticeship
500 C Street, N.W.
Suite 241
Washington, DC 20001

### Florida

Bureau of Apprenticeship
Division of Labor, Employment and Training
1320 Executive Center Drive
Suite 300
Tallahassee, FL 32399-0667

### Hawaii

Apprenticeship Division
Dept. of Labor and Industrial Relations
830 Punch Bowl Street
Honolulu, HI 96813

### Kansas

Apprenticeship Section
Division of Labor-Management Relations
    and Employment Standards
Kansas Department of Human Resources
401 SW Topeka, KS
Topeka, KS 66603-3182

## Kentucky

Apprenticeship and Training
Kentucky Labor Cabinet
Division of Employment Standards and Mediation
1049 U.S. 127 South
Frankfort, KY 40601

## Louisiana

Director of Apprenticeship
Department of Labor
P.O. Box 94094
Baton Rouge, LA 70804-9094

## Maine

Director of Apprenticeship
Bureau of Labor Standards
State House Station #55
Augusta, ME 04333

## Maryland

Apprenticeship and Training
Department of Economic and Employment Development
217 East Redwood Street
10th Floor
Baltimore, MD 21202

## Massachusetts

Division of Apprentice Training
Department of Labor and Industries
Leverett Saltonstall Building
Room 1107
100 Cambridge Street
Boston, MA 02202

## Minnesota

Division of Apprenticeship
Department of Labor and Industry
Space Center Building—4th Floor
443 Lafayette Road
St. Paul, MN 55101

## Montana

Apprenticeship and Training
Montana Department of Labor and Industry
P.O. Box 1728
Helena, MT 59620

## Nevada

Nevada Apprenticeship Council
Department of Labor
1445 Hot Springs Road
Suite 108
Carson City, NV 89710

## New Hampshire

New Hampshire Apprenticeship Council
Department of Labor
95 Pleasant Street
Concord, NH 03301-3593

## New Mexico

Director/Apprenticeship Section
New Mexico Department of Labor
501 Mountain Road, N.E.
Albuquerque, NM 87102

## New York

Director Employability Development
Department of Labor
The Campus Building, #12
Albany, NY 12240

## North Carolina

Director Apprenticeship Division
North Carolina Department of Labor
4 West Edenton Street
Raleigh, NC 27601

## Ohio

Ohio State Apprenticeship Council
Department of Industrial Relations
2323 West Fifth Avenue
Columbus, OH 43266-0567

## Oregon

Apprenticeship and Training Division
Oregon State Bureau of Labor and Industry
800 NE Oregon Street #32
Portland, OR 97232

## Pennsylvania

Apprenticeship and Training
Department of Labor and Industry
1303 Labor and Industry Building
7th and Forester Streets
Harrisburg, PA 17120

## Puerto Rico

Director, Incentive to the Private Sector Program
Right to Employment Administration
P.O. Box 4452
San Juan, PR 00936

## Rhode Island

Rhode Island Apprenticeship Programs
Department of Labor
220 Elmwood Avenue
Providence, RI 02907

## Vermont

Director
Vermont Apprenticeship Council
Department of Labor and Industry
National Life Insurance Drawer 20
Montpelier, VT 05602

## Virginia

Division of Apprenticeship Training
Department of Labor and Industry
13 South 13th Street
Richmond, VA 23219

## Virgin Islands

Division of Apprenticeship and Training
Department of Labor
#7 and #8 King Cross Street
Christiansted, Saint Croix
VI 00820

## Washington

Director, State of Washington
Department of Labor and Industries
P.O. Box 44500
Olympia, WA 98504-4500

## Wisconsin

Bureau Director
Department of Industry, Labor and Human Relations
P.O. Box 7972
Madison, WI 53707